IDEAS RELATING TO THE SCIENCES

VINCE FLYNT

Copyright © 2015, Vince Flynt

IDEAS RELATING TO THE SCIENCES

PREFACE

This book deals with various important topics in both the hard and soft sciences, namely, physics and economics in this instance. There are a number of unsolved problems or mysteries in these subjects and most of the materials here deal with them, suggesting the solutions.

The author has done considerable research on the topics in this book and here shares his thoughts with the reader, who may be able to develop further thoughts of his own.

Vince Flynt, Ph.D.

CONTENTS

1. The Very Nature Of Nature   5
2. Physics And Reality   7
3. The Controversies Of The Special Theory Of Relativity   9
4. The Velocity Of Light   19
5. More On Light   31
6. Space-Time   35
7. Role Of Consciousness In Physics   41
8. Gravity   43
9. More On Gravity   47
10. The Puzzle Of Quantum Entanglement   48
11. Unified Field Theory   51
12. Theory Of Everything (TOE)   58
13. More On The Theory Of Everything (TOE)   61
14. Scientific Method   63
15. The Important Soft Science: Economics   65
16. Economics As A Science For Solving Problems   66
17. Solutions In Economics   70
18. More Solutions In Economics   73
19. Conclusion   76
Bibliography   78

1 THE VERY NATURE OF NATURE

Scientists have often made use of highly abstract mathematics to describe nature. For example, relativity theory relies much on Riemannian geometry and tensor calculus, and, quantum theory relies on group theory. There is a saying that nature is mathematical. Mathematics deals with patterns and much of these patterns studied by mathematicians are also evident in nature. In looking for a solution to the Riemann Hypothesis, which is considered the most important unsolved problem in mathematics, some mathematicians are working on the hypothesis that perhaps the distribution of the zeros of the Riemann zeta function (which are mathematical objects) is similar to the distribution of quantum particles in the micro-world of particle physics. Group theory in mathematics is the study of the classification of members into groups based on their common properties and their characteristic possible permutations - it is the study of symmetries or patterns. Apparently, the symmetries or patterns which mathematicians deal with are also evident in nature, e.g., in the quantum world.

Is the similarity between some mathematical entities and some aspects of the natural world a mere coincidence? Or, is there a deeper meaning behind the similarity? It is apparent that mathematics is an effective instrument for the study of nature, and a well-known scientist had even dubbed this "the unreasonable effectiveness of mathematics", implying somewhat that the effectiveness is unexpected, surprising. Here perhaps we could infer that how we think, how a mathematician thinks, depends on and is influenced by our environment, which includes nature. When we deal with quantities, e.g., in algebraic manipulations, in geometrical computations (involving space), and in simple arithmetical calculations such as addition and subtraction, what kind of objects are we concerned with, if not the objects of nature? We cannot just simply say that we are dealing only with imaginary objects (objects which do not exist), for even to think of imaginary objects we need to experience nature first. It is nature, the environment, that provides the data for our thinking, our imagining, our mathematics. Our logical reasoning cannot exist in a vacuum or blank, where there is nothing, no feature, no data at all. The mathematician also cannot work in the blank.

Is nature really that simple? Does it indeed reveal itself so easily to the scientists and mathematicians? Are the apparent symmetries or patterns discerned in nature only a facade, beneath which lurks a deep, perhaps unfathomable, mystery? Scientists may, on the basis of such discernable patterns, jump to conclusions. They may experiment further and discern the same patterns. After that they may confirm their conclusions. But, the conclusions may be the result of one way of interpretation. However, there could be some other plausible interpretations. Scientific theories have often been modified to suit new evidences.

Symmetries or patterns in nature imply order, predictability. Scientists who interpret nature may be comparable to two persons looking at a cup which is half filled with water; one person looking at the cup may say it is quite empty, half empty, the other person may say it is quite full of water, half full of water. Thus, scientists may only be more conscious of the symmetries or orderliness of nature, unconsciously or consciously blocking out the asymmetries, the disorder, the chaotic. Apparently, the human mind prefers to see order, symmetry, which is normally equated with beauty; perhaps, the human mind has long been conditioned to react in this way; many of us have been taught to appreciate the beauty of nature since young. On the other hand, we should not forget the ugly aspects, the destructive power of nature, such as the harmful radiation of the sun's rays, the destructive lightning, the monstrous typhoons, the gigantic tidal waves, the terrible earth-quakes, the terrifying volcanic eruptions, the flash floods, et al. Incidentally, all these natural disasters tend to be unpredictable or hard to predict; they are the opposite of order, which is predictable. Such disastrous natural phenomena are normally regarded as chaos or turbulence, which is unpredictable, the outcome of which is unpredictable, wherein a slight change gives rise to a more than proportionate increase in the outcome, which is a characteristic known as nonlinearity.

Of course, symmetry, set pattern and order allow one to infer, arrive at logical conclusions, predict, forecast. On the contrary, with nonlinearity, disorder, disarray, chaos or turbulence, whereby no set pattern is discernable, all this will be quite impossible. Incidentally, chaos or turbulence is also one of the important outstanding problems in mathematics; a mathematical theory or solution which is capable of minimizing, if not eradicating, the undesirable or harmful effects of chaos or turbulence, e.g., by making its force and movements somewhat predictable to some extent, will benefit, e.g., the aircraft designer and the ship designer, who have to consider the possible effects of air turbulence and turbulence in the sea respectively, while carrying out their design work.

Therefore, in looking at and appreciating the symmetries and the beautiful patterns of nature, we should not forget its darker side, which is characterized by disorder, lack of pattern, lack of predictability, chaos or turbulence. To forget or ignore this darker side may be so at our peril.

2 PHYSICS AND REALITY

A number of points are brought up below for pondering and possible discussion. We should not regard our physics as the one and only reality. What is described below may seem like science fiction, but should not be brushed off as inconsequential.

Life And Existence
We take many things for granted. For example, we take it that all living things need oxygen and water to survive. We take it that planets are affected by gravity and must revolve round the sun. We take it that all living things have consciousness and intelligence. We take it that all living things propagate themselves and increase their population. We take it that all matter comprises of atoms. Et al. All this has become the standard reality in our minds; all this is life, existence.

Can anyone imagine a universe where all the above do not happen, that what we regard as life here in our planet have no semblance at all to any living thing that possibly exists in this universe? For example, there may be life without mind or consciousness (whereby life takes on a different meaning), life without the need to eat or respire whereby life may have an altogether different characteristic, life may even be just a vegetable-like existence (though possibly with consciousness), it may even be immortal. In other words, life or living beings may exist in this strange, vastly different universe but may be unrecognizable, even unseeable by our human sight or invisible. The life-forms there may be so different from ours that in our human minds they may not be regarded as life-forms at all. Surely, in a universe or environment where, for example, essential life-giving elements such as oxygen and water are non-existent, any life-forms, if they exist, must be vastly different from the life-forms here in our planet, so much so that they may not be regarded as life-forms by us mortals. In short, what we may see as barren in this vastly different universe may in fact be teeming with unrecognizable, perhaps "ghost-like", life-forms which may be invisible to our human eyes. This should not be a surprise. Even in this earthly existence of ours, some people and even animals, for example, dogs, are apparently capable of seeing "spiritual entities" or "ghosts" while the rest of their brotherhood live in this world in apparent blissful ignorance of the existence of such "frightful entities". It should be noted that the forces of nature in another universe may be vastly or even totally different from ours here, which probably account for the different life-forms that may exist there, that is, the physics there may be different.

Intelligence And Reasoning Power
Man has always prided himself on his high intelligence and reasoning power. We think we are so smart. But we still need to reason to arrive at the truths. That is, we need to have premises, facts, theorems, to reason with, we need to rack our brains. This reasoning process is often a time-consuming, even time-

wasting, process. There are probably intelligent beings in other universes who have some sort of telepathic or intuitive mind which is capable of knowing or arriving at the truths straight away without any use of our kind of reasoning process, thus saving plenty of time and mental energy. Can we then say that the reasoning human being is damn smart (as compared to these extra-terrestrials with the possibly telepathic, intuitive, lightning-quick mind)? These extra-terrestrials may know or understand straight away without any need of a reasoning process (unlike us mortals), as if by magic, a seeming miracle. All this represents a different reality.

Possible Danger Posed By Extra-Terrestrials
A few scientists have postulated that there is the possibility of Earth being attacked by extra-terrestrials or beings from other universes some day. Is Earth prepared to defend itself against these hostile aliens?

Conclusion
Is humanity prepared for the above-described possible new physics, new reality? Does anyone feel threatened by all this?

3 THE CONTROVERSIES OF THE SPECIAL THEORY OF RELATIVITY

Special Relativity Theory (SRT) has two postulates, one stating that the laws of physics are the same for all observers, and the other stating that the speed of light is the constant 186,000 miles per second, regardless of any reference frames. As a result of these postulates, SRT renders predictions such as: **1)** No object can travel faster than 186,000 miles per second (the speed of light itself); **2)** On approaching the speed of light, a moving object contracts in length in the direction of motion, while **3)** a clock traveling with the object slows down; **4)** The mass of an object multiplied by the square of the speed of light gives energy ($E = mc^2$); i.e., mass could be converted to energy and *vice versa*; **5)** Observers do not agree on the simultaneity of events - two events that are simultaneous for one observer might not be simultaneous for another.

There are evident inconsistencies among these predictions. There is also a philosophical problem relating to the nature of reality. Could there be more than one reality in Nature; that is, can reality be subjective, and only a matter of interpretation? This chapter explores the evident inconsistencies and the philosophical problem by developing arguments and providing numerical examples.

Introduction

Albert Einstein proposed the Special Theory of Relativity (SRT) in 1905 in his article "On the Electrodynamics of Moving Bodies". SRT has two postulates, as is stated in the Abstract above, from which is derived the motion of particles moving at close to the speed of light; it propounds the laws of motion for any particle. This does not mean Newton's theory was wrong. Newton's equations are contained within the relativistic equations, and are valid for velocities much less than the speed of light. For particles moving at low speeds, very much less than the speed of light, the differences between Einstein's laws of motion and those derived by Newton are very small. This is the reason why relativity does not play a large role in everyday life. SRT is now very well established as the description of motion of relativistic objects, i.e., those traveling at a significant fraction of the speed of light. It supersedes Newton's theory, but Newton's theory provides a very good approximation for objects moving at everyday speeds.

Illusion

SRT predicts that, upon approaching the speed of light, clocks slow down, moving objects contract in length in the direction of motion, and, a person's brain and bodily functions slow down. For example, a person on a moving vehicle (moving frame) that travels beside a beam of light (moving frame) in the same direction, and almost as fast as the beam of light (moving frame) itself, gauges the speed of the beam of light (moving frame). SRT predicts that this person on the moving vehicle (moving frame) traveling at almost the speed of the beam of light (moving frame) would find the speed of the beam of light (moving frame) to be unchanged, at 186,000 miles per second; this instead of the difference between the speed of the moving vehicle (moving frame) and the speed of the beam of light (moving frame), which would normally be the case.

This is claimed to be the case because, according to SRT, on the moving vehicle (moving frame) approaching the speed of light, the clock therein used to gauge the time traveled by the beam of light (moving frame) has slowed down by the same degree (say X %) as the ruler or measuring device therein used to gauge the distance traveled by the beam of light (moving frame) has contracted in length in the direction of the vehicle's motion (also X %), the greater the moving vehicle's traveling speed the more the clock slows down and the greater the length contraction of the ruler or measuring device.

All this is expressed in the following equation, which is in accordance with SRT:

$$\frac{186,000 \text{ miles} - X \text{ % of } 186,000 \text{ miles}}{1 \text{ second} - X \text{ % of } 1 \text{ second}} = 186,000 \text{ miles per second}$$

But (and this is a very important 'but'), if the moving vehicle's clock had not slowed down, and its ruler or measuring device had not contracted in length, (i.e., under normal conditions), then the speed of the light beam (moving frame), as gauged from the vehicle traveling besides it at almost the speed of light (moving frame), would have had been the difference between the speed of the light beam (moving frame) and the speed of the moving vehicle (moving frame); *e.g.*, 186,000 miles per second (speed of the light beam) minus 185,990 miles per second (speed of the moving vehicle), which is equal to 10 miles per second.

Thus, the speed of the light beam (moving frame), i.e., 186,000 miles per second, as gauged from the vehicle traveling besides it at almost the same speed (moving frame) in the same direction is evidently an *illusion*, which is somewhat similar to the situation whereby a driver in a car which is actually cruising at 60 miles per hour believes that his car is traveling at 30 miles per hour because the car's speedometer, which happens to be faulty, gives a reading of the car's cruising speed as 30 miles per hour instead of 60 miles per hour; i.e., the driver is misled by the car's faulty speedometer.

In the above-mentioned case, the person on the vehicle traveling at almost the speed of light (moving frame), say a space-ship, would not notice that his clock is ticking more slowly, time is passing more slowly for him, he is aging more slowly, and the length of the ruler or measuring device on his space-ship (moving frame) has contracted in the direction of motion (because there is nothing to compare with).

According to SRT, when this traveler on the space-ship traveling at almost the speed of light (moving frame) looks at a clock on Earth (stationary frame), he would perceive that the clock has slowed down, and when he looks at a ruler or measuring device on Earth (stationary frame) he would perceive that it has become shorter. But, this is evidently only an illusion, and not true, and, the clock ticking away on Earth (stationary frame) is actually ticking more quickly (which implies that time is passing more quickly) as compared to the traveler's clock on the space-ship traveling at almost the speed of light (moving frame), and the ruler or measuring device on Earth (stationary frame) is actually longer as compared to the traveler's ruler or measuring device on the space-ship traveling at almost the speed of light (moving frame) - according to SRT, the clock on the space-ship traveling at almost the speed of light (moving frame) has slowed down and both the length of the space-ship traveling at almost the speed of light (moving frame) and the length of the ruler or measuring device on this space-ship (moving frame) have contracted in length in the direction of motion, whilst the clock on Earth (stationary frame) has not slowed down and the ruler or measuring device on Earth (stationary frame) has not contracted in length. The person on Earth (stationary frame) would also notice the same things that the person on the vehicle traveling at almost the speed of light (moving frame) notices, i.e., the both of them notice the same things about one another.

The other thing each of them would agree on, which is important, is the constancy of light speed: 186,000 miles per second. Since neither of the two parties (on Earth (stationary frame) and on the space-ship traveling at almost the speed of light (moving frame)) had been aware that their respective clocks had been ticking away at different speeds, and, the lengths of their respective rulers or measuring devices had been different, each of them would have regarded the times shown by their respective clocks as the actual time and the lengths displayed by their respective rulers or measuring devices as the actual length, in which case there would be two sets of actual time and actual length, i.e., two sets of reality, a quite absurd situation.

As the third party looking on at the two cases described above and being aware of the circumstances, we could regard the time presented by the clock on Earth (stationary frame) as the actual or correct time and the time presented by the clock on the space-ship traveling at almost the speed of light (moving frame) as the distorted time. Also, as the third party who is all too familiar with the Special Theory of Relativity we could regard the length of the ruler or measuring device on Earth (stationary frame) as the actuality, and the length of the ruler or measuring device on the space-ship traveling at almost the speed of light (moving frame), which has contracted in the direction of the space-ship's motion, as the distorted length.

Inconsistency
We here consider the example of two space-ships traveling almost next to one another in the same direction, one (we call it X, which is a moving frame) traveling at almost the speed of light, say, 185,000

miles per second (as gauged from Earth, a stationary frame) and the other (we call it Y, which is another moving frame) traveling also at almost the speed of light, say, 185,500 miles per second (as gauged from Earth, the stationary frame). (Theoretically, no space-ship could travel at the speed of light - the Special Theory of Relativity posits that at the speed of light everything would be at a standstill - the mass of the space-ship would be infinite and the space-ship would not be able to accelerate anymore, the space-ship's length would have shrunk to zero and any clock within the space-ship would have stopped beating, registering zero time.) The speed of Y (moving frame) as gauged from X (moving frame) or vice versa is computed by using the following formula, as is posited by the Special Theory of Relativity:

$$v = \frac{b-a}{1 - ba/c^2}$$

where c = speed of light = 186,000 miles per second, b = speed of Y, = 185,500 miles per second (= 0.9973118c), a = speed of X, = 185,000 miles per second (= 0.9946236c)

$$\therefore v = \frac{0.9973118c - 0.9946236c}{1 - 0.9973118c \times 0.9946236c / c^2}$$

$$= \frac{0.0026882c}{1 - 0.9919498c^2 / c^2} = \frac{0.0026882c}{0.0080502}$$

$$= 0.3339295c \ (33.39295 \% \text{ of speed of light})$$

$$= 0.3339295 \times 186,000 \text{ miles per second}$$

$$= 62,110.887 \text{ miles per second}$$

$\therefore$ speed of Y as gauged from X = plus 62,110.887 miles per second (and not plus 500 miles per second (185,500 miles per second minus 185,000 miles per second), which should normally be the case - Y (moving frame) should appear to the traveler in X (moving frame) to be moving away from X (moving frame) in the same direction)

$\therefore$ speed of X as gauged from Y = minus 62,110.887 miles per second (and not minus 500 miles per second (minus (185,500 miles per second minus 185,000 miles per second)), which should normally be the case - X (moving frame) should appear to the traveler in Y (moving frame) to be moving away from Y (moving frame) in the opposite direction)

What would be X's and Y's respective speeds then (when gauged from the other), when X (moving frame) and Y (moving frame) travel in opposite directions (instead of the same direction)? The speed of Y (moving frame) as gauged from X (moving frame) and the speed of X (moving frame) as gauged from Y (moving frame) should each not exceed 186,000 miles per second, the speed of light, which represents the ultimate limit, the maximum possible speed any accelerating object could attain, as is postulated by the Special Theory of Relativity (and not respectively 370,500 miles per second (185,000 miles per second plus 185,500 miles per second), which should normally be the case), and, they are computed by using the following formula (which is described further on), which is in accordance with the Special Theory of Relativity:-

$$v = \frac{a+b}{1+ab/c^2}$$

where c = speed of light = 186,000 miles per second, a = speed of X = 185,000 miles per second (= 0.9946236c), b = speed of Y = 185,500 miles per second (= 0.9973118 c)
$\therefore$

$$v = \frac{0.9946236c + 0.9973118c}{1 + 0.9946236c \times 0.9973118c/c^2}$$
$$= \frac{1.9919354c}{1 + 0.9919498c^2/c^2} = \frac{1.9919354c}{1.9919498} = 0.9999927c$$
$$= 99.99927 \% \text{ of speed of light}$$
$$= 0.9999927 \times 186,000 \text{ miles per second}$$
$$= 185,998.64 \text{ miles per second}$$

$\therefore$ speed of Y as gauged from X = speed of X as gauged from Y = 185,998.64 miles per second (Y (moving frame) should appear to the traveler in X (moving frame) to be moving towards X (moving frame) in the opposite direction, and, X (moving frame) should appear to the traveler in Y (moving frame) to be moving towards Y (moving frame) in the opposite direction)

How strange and counter-intuitive it is to find the speeds of X (moving frame) and Y (moving frame) to be minus 62,110.887 miles per second and plus 62,110.887 miles per second respectively as gauged from

Y (moving frame) and X (moving frame) respectively (and not minus 500 miles per second and plus 500 miles per second respectively, which should normally be the case) in the first case above, and, to be each only 185,998.64 miles per second (less than the speed of light (186,000 miles per second) and not respectively 370,500 miles per second (185,000 miles per second plus 185,500 miles per second), which should normally be the case) in the second case above, one may think. Evidently, the respective clocks in X (moving frame) and Y (moving frame) were slowing down at different speeds and the respective rulers or measuring devices in X (moving frame) and Y (moving frame) were contracting in length to different extents, since the respective speeds of X (moving frame) and Y (moving frame) are different, *viz.*, 185,000 miles per second and 185,500 miles per second respectively (the higher the speed of the space-ship the more its clock would slow down and the more the length of its ruler or measuring device would contract).

As found by SRT, both the travelers in X (moving frame) and Y (moving frame) would each see the other's clock as being slower to the same degree and the other's ruler or measuring device as being shorter to the same degree. The dilemma here is to decide whether the clock on X (moving frame) or the clock on Y (moving frame) is giving the correct reading in time and whether the ruler or measuring device on X (moving frame) or the ruler or measuring device on Y (moving frame) is providing the correct measurement in the distance traveled/measured. It is evidently very difficult to decide thus. The travelers in X (moving frame) and Y (moving frame) would each naturally think that everything is fine and consider their respective gauging of the other's speed as correct (assuming that they have no knowledge at all about the Special Theory of Relativity).

However, if the travelers in X (moving frame) and Y (moving frame) noticed that the other's clock had been slower and the other's ruler or measuring devices had been shorter, they might each be puzzled and might each wonder whether whose clock and ruler or measuring device are accurate (assuming that they have no knowledge of the Special Theory of Relativity). Of course, if they had known the principles behind SRT, they would have realized that this phenomenon had been the result of "distortion" due to the creation of an intense gravitational field through travel at almost the speed of light, i.e., the slowing down of their respective clocks and the contraction in the lengths of their respective rulers or measuring devices are transient (they are not permanent - X's and Y's respective clocks would beat at the normal rate and the lengths of their respective rulers or measuring devices would return to their original length once the speeds of X (moving frame) and Y (moving frame) have returned from almost the speed of light (185,000 miles per second and 185,500 miles per second respectively) to the normal speeds, according to the Special Theory of Relativity).

The important question is if the spaceships', X's and Y's, times and length or distance measurements are "distorted" or not real, what should be the real time and real length or distance measurement? The other question, which is very important, that should strike a really sensible mind, is whether a ruler or measuring device, which is rigid and solid, could really contract in length and expand back to its original length in accordance with the Special Theory of Relativity, as is described above, as though it is made of rubber, which is flexible. But, to us, the third party looking on at these two scenarios, who have knowledge of the Special Theory of Relativity, both X's and Y's "real" times and "real" length or distance measurements are illusions and are indeed not real, and, the real time and real length or distance measurement would be those read off a clock and a ruler or measuring device on Earth (stationary frame), where the clock and the ruler or measuring device are free from the "distortional" effect of the intense gravitational field created through travel at almost the speed of light. All this is evidently a case of how we choose to interpret these three scenarios. A philosophical-minded person could choose the other interpretation, *viz.*, X's time and length or distance measurement are real to the traveler in X (moving frame), Y's time and length or distance measurement are real to the traveler in Y (moving frame), and, Earth's time and length or distance measurement are real to the resident on Earth (stationary frame), i.e., there are different realities. Should there be only one reality? Or should more than one reality be allowed?

We look at the first case pertaining to spaceships X (moving frame) and Y (moving frame) traveling at speeds of 185,000 miles per second (as gauged from Earth, a stationary frame) and 185,500 miles per second (as gauged from Earth, a stationary frame) respectively in the same direction almost next to one another. The speed of Y (moving frame) as gauged from X (moving frame) should normally be plus 500 miles per second (185,500 miles per second minus 185,000 miles per second) and the speed of X (moving frame) as gauged from Y (moving frame) should normally be minus 500 miles per second (minus (185,500 miles per second minus 185,000 miles per second)), as explained above. But, the speed of Y (moving frame) as gauged from X (moving frame) and the speed of X (moving frame) as gauged from Y (moving frame) should be plus 62,110.887 miles per second and minus 62,110.887 miles per second respectively, as computed by using the formula below, which is in accordance with SRT:

$$ v = \frac{b - a}{1 - ba/c^2} $$

where $c =$ speed of light $= 186,000$ miles per second, $b =$ speed of Y $= 185,500$ miles per second ($= 0.9973118c$), $a =$ speed of X $= 185,000$ miles per second ($= 0.9946236c$), so $v = 62,110.887$ miles per second. For instance, if the length of the ruler or measuring device on the space-ship contracts by 20 % while the space-ship travels at almost the speed of light, the relative speed of plus/minus 500 miles per second of each of the space-ships, X (moving frame) and Y (moving frame), should be recomputed/gauged as follows to produce the "distorted" speed of plus/minus 62,110.887 miles per second, which is in accordance with SRT:

$$\frac{62{,}110.887 \text{ miles}}{1 \text{ second}} = 400 \text{ miles (0.8 of 500 miles - due to ruler}$$

length contraction of **20 %**, 500 miles are gauged by space-ship traveler as 400 miles) ÷ 0.00644 second (due to clock on space-ship slowing down by **15528 %**, 1 second is gauged by space-ship traveler as 0.00644 second)

We now look at the second case pertaining to space-ships X (moving frame) and Y (moving frame) traveling at speeds of 185,000 miles per second (as gauged from Earth, a stationary frame) and 185,500 miles per second (as gauged from Earth, a stationary frame) respectively in opposite directions. The speed of Y (moving frame) as gauged from X (moving frame) and the speed of X (moving frame) as gauged from Y (moving frame) should each normally be 370,500 miles per second (185,000 miles per second plus 185,500 miles per second), as explained above. However, the speed of Y (moving frame) as gauged from X (moving frame) and the speed of X (moving frame) as gauged from Y (moving frame) should each not exceed 186,000 miles per second, the speed of light, which represents the ultimate limit, the maximum possible speed any accelerating object could attain, as found by SRT (and not respectively 370,500 miles per second (185,000 miles per second plus 185,500 miles per second)), which should normally be the case), and, they are computed by using the following formula, which is in accordance with SRT:-

$$v = \frac{a + b}{1 + ab / c^2}$$

where c = speed of light = 186,000 miles per second, a = speed of X = 185,000 miles per second (= 0.9946236c), b = speed of Y = 185,500 miles per second (= 0.9973118c) = 185,998.64 miles per second

For instance, if the length of the ruler or measuring device on the space-ship contracts by 60 % while the space-ship travels at almost the speed of light, the relative speed of 370,500 miles per second of each of the space-ships, X (moving frame) and Y (moving frame), should be recomputed/gauged as follows to produce the "distorted" speed of 185,998.64 miles per second, which is in accordance with SRT:

$$\frac{185,998.64 \text{ miles}}{1 \text{ second}} =$$

148,200 miles (0.4 of 370,500 miles - due to
length contraction of **60 %**, 370,500 miles are
gauged by space-ship traveler as 148,200 miles) ÷
0.79678 second (due to clock on space-ship slowing
down by **125.51 %**, 1 second is gauged by
space-ship traveler as 0.79678 second)

Thus, as is evident from the above examples, which are in accordance with SRT, to arrive at the two speeds, i.e., 62,110.887 miles per second and 185,998.64 miles per second, as well as other speeds, obtained by using the formulas of SRT,

$$v = (b-a) + (1 - ba/c^2) \text{ and } v = (a+b) + (1 + ab/c^2) \quad ,$$

the clocks and the rulers or measuring devices on the space-ships traveling at almost the speed of light would have to each respectively slow down and contract in length **at different rates** (and definitely not at the same rate). The only exception is evidently the case of the constancy of the speed of light, whereby the clock and the ruler or measuring device have to each respectively slow down and contract in length **at the same rate**, giving the **same percentage** decrease in the time gauged and the distance gauged, as follows, as found in SRT:

$$\frac{(186,000 \text{ miles } - X\% \text{ of } 186,000 \text{ miles})}{(1 \text{ second } - X\% \text{ of } 1 \text{ second})}$$
$$= 186,000 \text{ miles per second}$$

Why is the constancy of the speed of light the **exception**? Was it an adjustment or modification of the mathematics to 'ensure' the constancy of light speed? Could the speed of light not be variable, below, at, and above 186,000 miles per second, at various times, as some have suggested?

Conclusion
The important question pertaining to the above-described cases is if the spaceships', X's and Y's, times and length or distance measurements are "distorted" or not real, what should be the real time and real length or distance measurement?

However, to us, the third party looking on at these two scenarios, who have knowledge of the Special Theory of Relativity, both X's and Y's "real" times and "real" length or distance measurements are illusions and are indeed not real, and, the real time and real length or distance measurement would be those read off a clock and a ruler or measuring device on Earth (stationary frame), where the clock and the ruler or measuring device are free from the "distortional" effect of the intense gravitational field created through travel at almost the speed of light. All this is evidently a case of how we choose to interpret these three scenarios.

A philosophical-minded person could choose the other interpretation, *viz.*, X's time and length or distance measurement are real to the traveler in X (moving frame), Y's time and length or distance measurement are real to the traveler in Y (moving frame), and, Earth's time and length or distance measurement are real to the resident on Earth (stationary frame), i.e., there are different realities. Should there be only one reality? Or should more than one reality be allowed?

The other question, which is very important, that should strike a really sensible mind, is whether a ruler or measuring device, which is rigid and solid, could really contract in length and expand back to its original length in accordance with the Special Theory of Relativity, as is described above, as though it is made of rubber, which is flexible. Doesn't length contraction therefore appear absurd?

Another significant related point which should be noted is that (though experimental findings had confirmed that at high speeds, though very much less than the speed of light, clocks slow down) the contraction of rulers or measuring devices in the direction of motion at almost the speed of light is evidently only an inference, with no experimental basis. The point stated just above about the rigidity and solidity of the ruler or measuring device does imply that no experimental basis for such contraction could be expected to be forthcoming; the lack of experimental evidence could as a matter of fact be construed to mean that length contraction is actually an impossibility; even as an inference, it actually appears absurd, as is stated above. Why should people swear by a concept such as length contraction that is not backed by any experimental evidence at all, which is unscientific?

The above-described philosophical problem and inconsistencies, together with the evident lack of plausibility, and, empirical evidence, of length contraction, indeed cast doubt on the soundness of SRT, a shortcoming which should be seriously looked into.

4 THE VELOCITY OF LIGHT

The invariance of the velocity of light as is postulated by the Special Theory of Relativity had been confirmed by many experiments, e.g., the Michelson-Morley experiments. There are however a number of contradictions in Special Relativity. Quite a number of papers have been written about them.

The Special Theory of Relativity posits that a person on a moving vehicle, e.g., a very fast moving train (moving frame), traveling at close to the velocity of a beam of light (moving frame) in the same direction would find the velocity of the beam of light (moving frame) to be invariant at 186,000 miles per second, instead of the difference between the velocity of the very fast moving train (moving frame) and the velocity of the beam of light (moving frame), which would normally be the case. This is because, according to the Special Theory, on the very fast moving train (moving frame) approaching the velocity of light the clock therein used to gauge the time traveled by the beam of light (moving frame) has slowed down by the same degree as the ruler therein used to gauge the distance traveled by the beam of light (moving frame) has contracted in length in the direction of the very fast moving train's motion, the greater the very fast moving train's traveling velocity the more the clock slows down and the greater the length contraction of the ruler. This is expressed in the following equation (the velocity of the beam of light (moving frame) being the distance it traveled divided by the time it took to travel this distance), which is in accordance with the Special Theory of Relativity:-

$$(186{,}000 \text{ miles} - \beta\% \text{ of } 186{,}000 \text{ miles})/(1 \text{ second} - \beta\% \text{ of } 1 \text{ second}) = 186{,}000 \text{ miles per second}$$

In other words, there has to be a same percentage decrease in the time gauged and the distance gauged due to the respective slowing down of the clock and contracting in length of the ruler therein the very fast moving train (moving frame), in order for the velocity of the beam of light (moving frame) to remain invariant, which is consistent with mathematical logic - this condition has to apply in order for the velocity of the beam of light (moving frame) to remain invariant. It should be noted here that having the same percentage decrease in the time gauged and the distance gauged, as is shown in the above equation, seems an improbable occurrence.

There is however something rather unusual related to the above concept. According to the Special Theory of Relativity, the person on the very fast moving train traveling at close to the velocity of light (moving frame) gauging the velocity of the beam of light traveling in the same direction (moving frame) would not notice that the clock on his very fast moving train (moving frame) has slowed down and the ruler therein has contracted in length in the direction of motion. In other words, everything would appear normal to him, despite the fact that his clock has actually slowed down and his ruler has actually contracted in length in the direction of motion, as is postulated by the Special Theory of Relativity. But, according to the Special Theory of Relativity, when he (moving frame) compares himself to a person on the ground who is not moving (stationary frame), he could even consider himself stationary (stationary frame) while thinking that the person on the ground (who is not moving) is actually moving (moving frame), i.e., all movements are relative. He (moving frame) would notice that the clock on the ground (stationary frame) is slower and the ruler on the ground (stationary frame) is shorter. The person on the ground who is not moving (stationary frame) would also notice that the clock on the very fast moving train (moving frame) is slower, the ruler therein is shorter, and, the length of the very fast moving train (moving frame) is shorter. In other words, both the train-traveler (moving frame) and the person on the ground who is not moving (stationary frame) would notice that the other's clock is slower and the other's ruler is shorter, and, according to the Special Theory of Relativity, the slowing down of clocks and the shortening of rulers would appear to be by the same degree for both.

But, it is actually the clock on the very fast moving train traveling at close to the velocity of the beam of light in the same direction (moving frame) which has slowed down and the ruler therein which has contracted in length (in the direction of motion) as is postulated by the Special Theory of Relativity, and not those on the ground (stationary frame). To the traveler on the very fast moving train (moving frame) who is gauging the velocity of the beam of light traveling in the same direction (moving frame), the beam of light (moving frame) appears to take less time (time dilation) to travel a shorter distance (length contraction), which, according to the Special Theory of Relativity and in accordance with the following equation, explains the invariance of the velocity of light at all inertial frames:-

$$(186{,}000 \text{ miles} - \beta\% \text{ of } 186{,}000 \text{ miles})/(1 \text{ second} - \beta\% \text{ of } 1 \text{ second}) = 186{,}000 \text{ miles per second}$$

The velocity of the beam of light (moving frame) is obtained by dividing the distance traveled by the beam of light (as gauged by the ruler on the very fast moving train traveling at close to the velocity of light - moving frame) by the time it took to travel that distance by the beam of light (as gauged by the clock therein the very fast moving train - moving frame), the gauging being carried out by the traveler on the very fast moving train (moving frame). In the above example, the clock on the very fast moving train traveling at close to the velocity of light (moving frame) slows down and gauges a slower time, $\beta\%$ slower. The ruler therein also contracts in length in the direction of motion by $\beta\%$; however because of this it should gauge any object as "longer" due to a change in the scale of the ruler on contracting in length in the direction of motion (refer to Appendix at the back), and, this evidently gives rise to an inconsistency when computing the velocity of the beam of light (moving frame). This inconsistency in the computation of the velocity of light is described in the example below.

The person on the very fast moving train (moving frame) traveling at close to the velocity of the beam of light (moving frame) in the same direction could be considered one inertial frame, and, the ground level (stationary frame) could be considered another inertial frame. If, traveling at close to the velocity of light, both his clock has slowed down and his ruler has contracted in length in the direction of motion by the same degree (β %), the beam of light (moving frame) whose velocity he is gauging, according to the Special Theory of Relativity, would also appear to him to have traveled a distance (from one designated point (Point A) to another designated point (Point B) - reference, stationary frame) which is shortened by the same degree (β %), i.e., the distance between the two designated points Point A & Point B (reference, stationary frame) traveled by the beam of light appears to him to have shortened by the same degree (β %). We would describe below what happens when the train-traveler traveling at close to the velocity of light gauges the velocity of the beam of light traveling alongside his train in the same direction:-

```
reference        l------------------------------------------------------l
frame            A                                                      B

beam of light    ------------------------------------------------------->

train-traveler/  ----------------------------------->
train
```

(1) We first recapitulate and make clear the following conditions which are stipulated by the Special Theory of Relativity:

(a) Clocks slow down and rulers contract in length in the direction of motion, at close to the velocity of light, but these would not be noticed by the person traveling at close to the velocity of light (moving frame) who possesses such clocks and rulers. These do not apply to the clock and ruler of the person on the ground who is not moving (stationary frame).

(b) The train-traveler traveling at close to the velocity of light (moving frame) and the person on the ground who is not moving (stationary frame) would notice one another's clock and ruler as respectively slower and shorter than his own. Both of them would each notice the other's clock slowing down and ruler length shortening to the same degree (β % for the clock and ruler of the train-traveler traveling at close to the velocity of light (moving frame) and also β % for the clock and ruler of the person on the ground who is not moving (stationary frame)). Though the train-traveler's clock has slowed down and his ruler has contracted in length in the direction of motion while he is traveling at close to the velocity of light, as is described in (a) above, he would not notice these happenings. All this does not apply to the clock and ruler (which are not affected by (a) above) of the person on the ground who is not moving (stationary frame).

(c) All movements are relative. When the train-traveler traveling at close to the velocity of light (moving frame) compares himself to a person on the ground who is not moving (stationary frame), he could consider himself stationary (stationary frame) while thinking that the person on the ground (who is not moving) is actually moving (moving frame) - both parties could each regard themselves as stationary (stationary frame) and consider the other party in motion (moving frame).

(d) The train-traveler traveling at close to the velocity of light (moving frame) would see the distance from Point A to Point B at the embankment besides the railway tracks (reference, stationary frame - you might substitute this distance with a ruler) traveled by the beam of light whose velocity he is gauging as having shortened. This shortening (by β %) of the distance from Point A to point B (reference, stationary frame) would be by the same degree as his own clock has slowed down (β %) and his own ruler has contracted in length in the direction of motion (β %), both of which he does not notice. This point is the equivalent of Point (b) above.

(e) The velocity of the beam of light (moving frame) is obtained by dividing the distance traveled by the beam of light (as gauged by the ruler on the very fast moving train traveling at close to the velocity of light - moving frame) by the time it took to travel that distance by the beam of light (as gauged by the clock therein the very fast moving train - moving frame), the gauging being carried out by the traveler on the very fast moving train (moving frame), the greater the very fast moving train's traveling velocity the more the clock slows down and the greater the length contraction of the ruler. For the velocity of the beam of light (moving frame) to be measured as invariant, the clock in the train traveling at close to the velocity of light (moving frame) used to gauge the time traveled by the beam of light (moving frame) has to gauge a time which has slowed down by the same degree (β %) as the distance traveled by the beam of light (moving frame) which has shortened (also β %) as is gauged by the ruler therein.

(2) The following is what happens when the train-traveler's clock and ruler slows down and contracts in length in the direction of motion respectively by 50 % each and the distance from Point A to Point B (reference, stationary frame) traveled by the beam of light whose velocity the train-traveler is gauging has shortened by 50 % in his eyes:

(a) Before the distance from Point A to Point B (reference, stationary frame) traveled by the beam of light has shortened by 50 % (as is gauged at the ground level - stationary frame):

  (i) Distance from Point A to Point B (reference, stationary frame) = 1 meter
  (ii) Time taken by the beam of light (moving frame) to travel this 1 meter = x second, i.e., velocity of the beam of light (moving frame) = 1 meter per x second

The above distance of 1 meter in (i) is gauged at the ground level (stationary frame) with a ruler which has not contracted in length (by 50 %, as yet). The above time of x second in (ii) is gauged at the ground level (stationary frame) with a clock which has not slowed down (by 50 %, as yet).

(b) (i) After the ruler on the moving train (moving frame) has contracted in length in the direction of motion by 50 %, it would gauge the above-mentioned 1 meter (distance from Point A to Point B (reference, stationary frame), before shortening) traveled by the beam of light as 2 meters. (A shortened ruler gauges an object as "longer" while a lengthened ruler gauges an object as "shorter", due to the different scales, as is explained in the Appendix.)

(ii) After the clock on the moving train (moving frame) has slowed down by 50 %, at the same time that the ruler has contracted in length by 50 %, it would gauge the above-mentioned x second taken to travel this measured distance of 2 meters as 1/2 x second.

(iii) That is, the beam of light (moving frame) is now gauged from the moving train (moving frame) as requiring 1/2 x second to travel the distance of 2 meters (before the shortening of the distance between Point A & Point B (we assume here that the distance between Point A & Point B has not shortened) - in accordance with the Special Theory of Relativity this distance would also be shortened (also by 50 %) as seen by the train-traveler, as is described below); this translates into a velocity of 4 meters per x second, or, 4 times velocity of light.

(c) After the distance from Point A to Point B (reference, stationary frame) traveled by the beam of light has shortened by 50 %, as is seen by the train-traveler traveling at close to the velocity of light (moving frame), in accordance with the Special Theory of Relativity:

(i) The 1-meter ruler which has contracted in length in the direction of motion by 50 % would gauge the above-mentioned 50 % shortened distance from Point A to Point B (reference, stationary frame) as 50 % shorter than before (i.e., 0.5 meter instead of 1 meter), and would read "0.5 meter" (instead of "1 meter" - full length of the 1-meter ruler) on its 50 % shortened length, which is in accordance with the Special Theory of Relativity. (The ruler now has a different scale as compared to its scale before length contraction. Refer to Appendix for explanation.)

(ii) When the scale of the ruler has changed due to the ruler's length contraction (by 50%), and the clock has slowed down (by 50 %), the beam of light (moving frame) being gauged as requiring 1/2 x second to travel the distance of 2 meters, as is described in (b) (ii) above, the time taken by the beam of light (moving frame) to travel 0.5 meter now = (0.5 meter ÷ 2 meters) x 1/2 x second = 1/8 x second, i.e., the velocity of the beam of light (moving frame) is now 0.5 meter per 1/8 x second, or, 4 meters per x second/4 times velocity of light, which is an inconsistency (the velocity of the beam of light should have remained 1 meter per x second/1 time velocity of light, as is postulated by the Special Theory of Relativity).

The computations above have been carried out in accordance with the conditions stipulated by the Special Theory of Relativity which are described in (1) above. The above inconsistency is evidently due to the change of the scale of the ruler which has contracted in length in the direction of motion by 50 %. It requires attention.

In order for the velocity of the beam of light to remain/appear invariant in this instance, i.e., remain at 1 meter per x second, one of the following has to happen:-

1) When the clock slows down by 50 %, the ruler should increase in length by 100 %.
2) When the ruler decreases in length by 50 %, the clock should quicken by 100 %.
3) When the clock slows down by 50 % and the ruler decreases in length by 50 %, the beam of light (moving frame) should slow down by 400 %.

How do we explain the invariance of the velocity of light at all inertial frames if as described above there is an inconsistency relating to length contraction? The postulates of the Special Theory of Relativity evidently imply that the invariance of the velocity of light at all inertial frames is only an illusion - if the velocity of light were to *appear* invariant, according to the Theory, lengths have to contract (Lorentz contraction) and clocks have to slow down (time dilation), at the same rate, while moving at close to the velocity of light. We here ask the important question: If lengths do not contract and clocks do not slow down at close to the velocity of light, as are postulated by the Theory, would the velocity of light still appear invariant? In all this, we should also bear in mind that while the slowing down of clocks (time dilation) when traveling at high velocities is an experimentally proven phenomenon length contraction (Lorentz contraction) has not been experimentally proven and remains an inference. In view of the above-mentioned inconsistency, there could be another explanation or reason for the invariance of the velocity of light at all inertial frames, e.g., length expansion, as is described above, or, some other valid reasons. We have to find a fool-proof reason or reasons to explain why the velocity of light always appears invariant at all inertial frames, which would corroborate the experimental evidence that the velocity of light is invariant.

There is an evident way out of this difficulty or dilemma, which would be described here. Let us here recapitulate the important point brought up above, which is as follows:-

In order for the velocity of the beam of light to remain/appear invariant, i.e., remain at 1 meter per x second, one of the following has to happen:

1) When the clock slows down (time dilation) by x %, the ruler should increase in length (length expansion) by y %.

2) When the ruler decreases in length (length contraction) by x %, the clock should quicken (time contraction) by y %.

3) When the clock slows down (time dilation) by q % and the ruler decreases in length (length contraction) by q %, the beam of light (moving frame) should slow down by r %.

We would add a fourth option to the above three options, which is as follows:

4) When the clock slows down (time dilation) by q % and the ruler remains the same in length (unchanged in length) if Lorentz contraction were not an actuality and does not happen, the beam of light (moving frame) should slow down by s %. (s % < r %)

Since the phenomenon of clocks slowing down (time dilation) while traveling at high velocities had been confirmed by experiments and length contraction has not been confirmed experimentally as yet but is an inference only, Item (2) above (which states clock quickening, an unproven phenomenon and the reverse and contradiction of the experimentally proven clock slowing down phenomenon (time dilation)) could be ruled out, while Items (1), (3) and (4) are possibilities, however remote these possibilities might be. Though the intense gravitational field caused by travel at almost the velocity of light might account for the slowing down of clocks (for which experimental evidence had already been obtained as is stated above) and therefore time, as well as the brain and bodily functions of a person, it evidently hardly suffices as an explanation for length contraction (for which experimental evidence has yet to be found).

We should remember that length contraction is after all an unconfirmed inference (unlike time dilation which had been proven by experiments as is stated above). The same would apply to length expansion. There is probably no such things as length contraction or length expansion. It is difficult to envision or imagine a rigid object such as a ruler or meter rod contracting in length or expanding in length as though it is made of rubber, which is flexible, and such a phenomenon should be regarded as improbable; length contraction and length expansion could therefore be regarded as only illusions at most, more apparent than real. Because of this, Items (1) and (3) above would appear remotely probable with Item (2) completely ruled out as is stated above, while Item (4) is most probable. But Item (4) above implies that the velocity of light would appear to exceed the 186,000 miles per second limit (the slowed down clock ("time dilated" clock) and the ruler which remains the same in length (does not contract in length) would now together gauge the velocity of the beam of light (which is actually 186,000 miles per second) as more than 186,000 miles per second - as the clock has slowed down (time dilation), the beam of light would now (appear to) take less time to travel the same distance, i.e., the velocity of the beam of light now appears to be greater, this higher velocity being determined by dividing the distance traveled by the time taken to travel this distance), 186,000 miles per second being the limit of the velocity of light which is postulated by the

Special Theory of Relativity - the velocity of light could never exceed this limit as is postulated by the Theory. Thus, the above is evidently an *illusion* caused by the slowing down of the clock while the length of the ruler remains unchanged (does not contract), both the clock and the ruler having been utilized to gauge the velocity of the beam of light. That is, Item (4) above would produce the *illusion* of the beam of light (which has an actual velocity of 186,000 miles per second) having a velocity of more than 186,000 miles per second. All this would be another "headache" for the Special Theory of Relativity, which states that no moving object including light could exceed the velocity limit of 186,000 miles per second. As is stated in Item (4) above, in order for the slowed down clock (slowed down by q % for example) and the ruler whose length has not contracted but remains the same to gauge the velocity of the beam of light as invariant (invariant at 186,000 miles per second), the actual velocity of the beam of light has to be less than 186,000 miles per second (the beam of light should slow down by s %, as is stated in Item (4) above); this would of course result in the *illusion* that the velocity of the beam of light is invariant (unchanged at 186,000 miles per second) while the actual velocity of the beam of light is less than 186,000 miles per second - if the clock used to gauge the velocity of the beam of light (which is actually less than 186,000 miles per second, say d miles per second) had not slowed down (time dilation) but remained ticking at the same rate, the velocity measured would certainly be less than 186,000 miles per second (which is as stated just above the actual velocity, i.e., d miles per second).

However, of the four options above, Items (1), (2), (3) and (4), Item (4) is hence evidently the most realistic and probable. We recapitulate here: Item (4) states that there is time dilation but no length contraction, i.e., clocks would slow down at high velocities, e.g., velocities close to the velocity of light, but at such high velocities rulers would not contract in length in the direction of motion and would remain the same in length. Based on these conditions of Item (4), there is a logical, more sensible explanation for the invariance of the velocity of light, which would be described in the following:-

A person on a moving vehicle, e.g., a very fast moving train (moving frame), traveling at close to the velocity of a beam of light (moving frame) in the same direction would find the velocity of the beam of light (moving frame) to be invariant at 186,000 miles per second, instead of the difference between the velocity of the very fast moving train (moving frame) and the velocity of the beam of light (moving frame), which would normally be the case. This is because, according to the Special Theory of Relativity, on the very fast moving train (moving frame) approaching the velocity of light the clock therein used to gauge the time traveled by the beam of light (moving frame) has slowed down by the same degree (say β %) as the ruler or measuring device (stated as meter stick or measuring rod in some texts) therein used to gauge the distance traveled by the beam of light (moving frame) has contracted in length in the direction of the very fast moving train's motion (also β %), the greater the very fast moving train's traveling velocity the more the clock slows down and the greater the length contraction of the ruler or measuring device. This is expressed in the following equation (the velocity of the beam of light (moving frame) being the distance it traveled divided by the time it took to travel this distance), which is in accordance with the Special Theory of Relativity:

(186,000 miles − β % of 186,000 miles)/(1 second − β % of 1 second) = 186,000 miles per second

The explanation in this case for the invariance of the velocity of light is based on the conditions stipulated in Item (3) above, wherein length contraction is evidently not so probable, as is explained above. Also, to have length contraction and time dilation happen to the same degree (shown in the above equation as β % each) is not that probable. We would re-construe this case using the more realistic and probable conditions stated in Item (4) above, namely, time dilation, which is a proven phenomenon, and absence of length contraction, which is to be expected. We would use a simple diagram to explain, which is as follows:

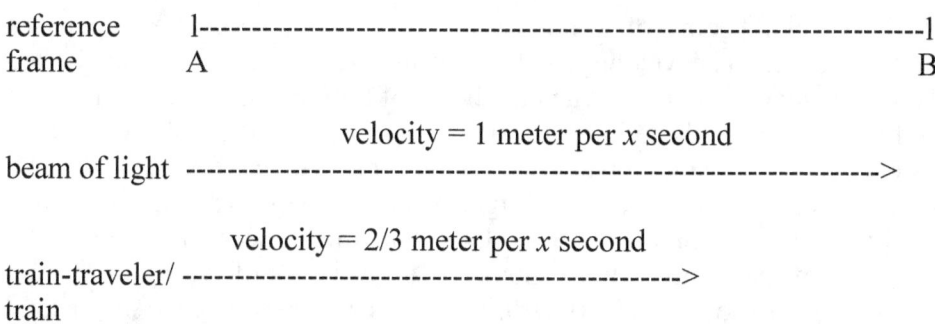

The above train-traveler traveling at two-third the velocity of light besides a beam of light in the same direction is gauging the velocity of the beam of light. The velocity of the beam of light is 1 meter per x second while the velocity of the train, being two-third the velocity of light, is 2/3 meter per x second. Under normal circumstances, the train-traveler would gauge the velocity of the beam of light traveling in the same direction besides his train as 1/3 meter per x second, obtained by deducting the velocity of the train from the velocity of the beam of light, i.e., 1 meter per x second minus 2/3 meter per x second. But, the train-traveler is now experiencing some abnormal conditions while traveling at two-third the velocity of light, i.e., 2/3 meter per x second, as, by Item (4) above, his clock slows down while his ruler or measuring rod remains unchanged in length (does not experience length contraction). Say, e.g., the train-traveler's clock has slowed down by two-third while his ruler or measuring rod remains the same in length, while traveling at two-third the velocity of light, or, 2/3 meter per *x* second. The train-traveler's unchanged ruler or measuring rod would now gauge 1/3 meter (the velocity of the beam of light should be 1/3 meter per *x* second under normal circumstances as is stated above) as 1/3 meter still but his clock which has slowed down by two-third would now gauge the time taken to travel the distance of 1/3 meter as 1/3 *x* second (and not x second), i.e., the train-traveler would now gauge the velocity of the beam of light traveling in the same direction besides his train as 1/3 meter per 1/3 *x* second, which is the same as 1 meter per x second, which is the velocity of light! Thus, to the train-traveler, the velocity of the beam of light traveling in the same direction besides his train is invariant, i.e., still 1 meter per *x* second, instead of 1/3 meter per *x* second. Therefore, the conditions of Item (4) above, namely time dilation and absence of length contraction, could be incorporated into a revised Special Theory of Relativity, whereby the inconsistency described above would be gone.

Thus, with length contraction (Lorentz contraction) out of the equation and with time dilation still playing a role, the above-described inconsistency evidently does not arise.

The other reason put forward by the Special Theory of Relativity for the invariance of the velocity of light is that light is unaffected by its source. The velocity of light would be invariant if gauged from different inertial frames, e.g., when the velocity is gauged when the beam of light is emitted from a source which is stationary (not moving), for instance, the headlight of a stationary car, and, when the velocity is gauged when the beam of light is emitted from a source which is moving, for instance, the headlight of a moving car - in both these instances the velocity of the beam of light would be the same, as is postulated by the Special Theory of Relativity, though common sense dictates that in the second instance, the instance of the moving car, the velocity of the beam of light should be the velocity of the beam of light (186,000 miles per second) plus the velocity of the car (say 0.014 miles per second), giving a total velocity of 186,000.014 miles per second. The answer to this abnormality, according to the Special Theory of Relativity, is that the beam of light is independent of its source, the car headlight, and is not affected by this source. This implies that if the car were to travel at a higher velocity than the velocity of the beam of light the car would be moving in front of the beam of light, while the beam of light would be tagging behind the car.

However, by the many experimental results, e.g., the Michelson-Morley experiments, the velocity of light should be invariant, but, a fool-proof theory is needed to explain it.

APPENDIX

We here consider a map of size p feet in length by q feet in breadth with a scale of 1 is to 100,000 (1 inch on the map represents 100,000 inches on the actual ground) which shows only a portion of our globe. To have the whole globe represented in this map p feet in length by q feet in breadth, it is evident that we have to change its scale, e.g., change its scale to 1 is to 1,000,000 (1 inch on the map represents 1,000,000 inches on the actual ground). The above-mentioned ruler which has contracted in length is analogous to a map whose scale has been changed to allow it to represent a larger area (i.e., the whole globe), e.g., from 1 is to 100,000 to 1 is to 1,000,000 - with the new scale of 1 is to 1,000,000, 1 inch on the map now represents a length of 1,000,000 inches instead of 100,000 inches on the actual ground. If we change the scale of the above-mentioned map which shows only a portion of our globe to 1 is to 1,000,000 from 1 is to 100,000, we would have a much smaller map showing the same portion of our globe, whose dimensions would be 1/10 p feet in length by 1/10 q feet in breadth, a contracted map which is analogous to the above-mentioned ruler which has contracted in length. A 1-meter-long ruler which has contracted in length by 50 %, e.g., would now gauge the length of 1 meter as 2 meters (and not 0.5 meter in accordance with the Special Theory of Relativity), and this is evidently the cause of the above-described inconsistency relating to the computation of the velocity of light. In fact, for the 1-meter-long ruler to gauge the length of 1 meter as 0.5 meter it would have to increase in length by 100 %. A shortened ruler would gauge an object as "longer" while a lengthened ruler would gauge an object as "shorter".

Let us look at a simple example here. For instance, the 1-meter-long ruler used to gauge distance has contracted in length in the direction of motion by 20 %. The clock used to gauge time, which has also slowed down by 20 %, according to the Special Theory of Relativity, would now gauge the time taken, say t, to travel the distance between two designated points (reference, stationary frame), say u, as having decreased by 20 % to become 0.8 t. Though the 1-meter-long ruler, which has contracted in length by 20 %, still reads "1 meter" in length, it is in effect shorter by 20 % (actually only 0.8 meter in length). Therefore, when it gauges the distance traveled, u, above, this distance u would now be gauged as 1.25 u, and not 0.8 u in accordance with the Special Theory of Relativity. As stated above, the Special Theory of Relativity theorizes that for a beam of light (moving frame) to remain invariant in velocity the beam of light (moving frame) has to take less time (time dilation) to travel a shorter distance (length contraction) - in effect, β % less time to travel a distance shorter by β %, in accordance with the following equation, which implies that the velocity of the beam of light (moving frame) would remain invariant, e.g., 0.8 t (time) to travel 0.8 u (distance) after "time dilation" and "length contraction":-

(186,000 miles − β % of 186,000 miles)/(1 second − β % of 1 second) = 186,000 miles per second

But, as explained above, this would not be the case; the beam of light (moving frame) would have been gauged as having taken $0.8\ t$ (time) to travel $1.25\ u$ (distance). This is an inconsistency in the Special Theory of Relativity.

5 MORE ON LIGHT

Does light really need a medium, which is termed the luminiferous ether, for its transmission as is in the case of sound which requires a fluid such as air or liquid as the medium for its transmission? How about the constancy of the velocity of light?

Position Of Luminiferous Ether

Einstein in his relativity theory had done away with the luminiferous ether being the medium for the transmission of light. But a number of scientists today are apparently trying to resurrect the role of the luminiferous ether. Is the luminiferous ether indeed necessary for the transmission of light?

Light should not be expected to act the same as sound as they are both intrinsically different. Sound is the result of vibrations in the air (a medium of transmission for sound), vibrations which reach our ear-drums at varying frequencies, i.e., sound is actually moving, vibrating air which affects our ear-drums. Light has been described as a "wave/particle" object - it could be viewed as both a wave and a particle. Does it make sense to expect light too to travel through a medium like sound does? The quantum particle of light is the photon, which is rather similar to the other quantum particles such as the positron, electron, proton and neutron, etc., whereas sound has no quantum particle within it, being just vibrations or physical movements of the air (which affect the sensitive ear-drums). Quantum particles evidently do not need a medium for their functionality. Why should quantum light need a medium such as the luminiferous ether for its transmission?

The Michelson-Morley experiment and others did not find any evidence of this medium, the luminiferous ether. And, apparently from this "null" result of the Michelson-Morley experiment it had been concluded, evidently without much justification, that the velocity of light is invariable, and that nothing could travel faster than the velocity of light, which has apparently just recently been disproved by CERN. Moreover, quantum particles have been found to be capable of teleportation, i.e., transport to another location in space instantaneously, to display "weirdness" (i.e., appear strange and incomprehensible). From all this, it could be concluded that unlike sound, which requires a medium such as air or liquid for its transmission, for light such a medium is not a necessity.

On the other hand, if the luminiferous ether exists (as the medium for the transmission of light), doesn't it have to be composed of atoms, or, quantum particles, as well (the luminiferous ether here appears comparable, e.g., to the carrier signal which carries the picture and sound signals to our TV set all the way from the transmitter at the broadcasting station which could be many miles away)? For those who have been toying with the idea of the luminiferous ether or who are convinced that it exists, how would they describe this medium, e.g., what they are made of, whether they are comprised of atoms, etc.?

Tests

The only way to determine that the luminiferous ether does indeed exist is to physically detect it through experiments, which had evidently so far not been successful, instead of theorisation, and, since the experiments so far had not detected it, it could be concluded that the luminiferous ether does not exist (provided that the experiments had not been faulty).

Scrutiny

However, the following important questions should be lavished with some consideration: How is light able to travel very long distances at the very great velocity of 186,000 miles per second in a vacuum? Is some very strong force, a yet unknown, undetected, force, perhaps, a very strong, very high frequency, carrier wave (similar to the above-mentioned carrier signal), which might be interpreted by some as the medium of transmission, carrying light along? Or, is the movement of light entirely self-propagating, i.e., without the aid of any external force? Shouldn't the velocity of light be at, above and below 186,000 miles per second at various points in time, i.e., be variable, as clocks and watches, in fact all mechanisms, natural and artificial, do go faster or slower to varying degrees at various times, which explains why all equipment, including precision equipment, have to be calibrated from time to time in order that accuracy at the accepted level is maintained?

Superluminal Motion

A person could be deceived about the time by a not perfect clock, and, the distance by an also not perfect measuring rod. All this implies that time is subjective and not absolute, or, as Einstein had put it, relative, depending on the situation.

There should be a sufficient reason to explain why the clock, and, the brain and bodily functions of the person slow down, and the length of the measuring rod contracts in the direction of motion, on approaching the velocity of light, while at the velocity of light the mechanism of the clock and time are at a standstill and the length of the measuring rod is zero, which is important. Though the intense gravitational field caused by travel at almost the velocity of light might account for the slowing down of the clock (for which experimental evidence had been obtained) and therefore time, as well as the brain and bodily functions of a person, it apparently hardly suffices as an explanation for the contraction of the length of the measuring rod in the direction of travel at almost the velocity of light (for which experimental evidence has yet to be found). Though the constancy of the velocity of light as gauged from the Earth is apparently a well-proven phenomenon, no one has yet been able to travel at almost the velocity of light and gauge the velocity of a light beam by traveling besides it in the same direction though it has been postulated that in this instance the velocity of the light beam would remain the same. Despite the experimental findings that at high velocities, though very much less than the velocity of light, clocks slow down, the contraction of measuring rods in the direction of motion at almost the velocity of light is apparently only an inference, with no experimental basis.

The following equation describes how the velocity of light (v) is derived:-

$$v = d/t,$$

where d represents the distance traveled by the light beam and t represents the time taken by the light beam to travel the distance d

Since time is relative (and not absolute) and depends on the mechanism of the clock, which slows down on approaching the velocity of light, it could be arbitrary. The clock which is used to gauge the time t taken by the light beam to travel the distance d might not slow down uniformly (at the same rate) on approaching the velocity of light (under normal, earthly conditions time varies from clock to clock by minutes or more - there is apparently some uncertainty in the mechanism of clocks). Besides, the measuring rod used to gauge the distance d traveled by the light beam in time t might not contract in length uniformly (at the same rate) on approaching the velocity of light. If the clock does not slow down uniformly (at the same rate) and the measuring rod does not contract in length uniformly (at the same rate) on approaching the velocity of light there is the probability that the velocity of light (v) as represented by d/t would be variable, higher than 186,000 miles per second at times, below 186,000 miles per second at other times, or, equal to 186,000 miles per second at yet other times. Moreover, in accordance with the Special Theory of Relativity, for the velocity of light to really remain constant, on approaching the velocity of light the clock must slow down to the same degree as the contraction in the length of the measuring rod. We describe these possible outcomes as follows:-

$$\text{(i)} \quad S^\% > C^\% \rightarrow I^l$$
$$\text{(ii)} \quad S^\% < C^\% \rightarrow D^l$$
$$\text{(iii)} \quad S^\% = C^\% \rightarrow S^l$$

where $S^\%$ represents percentage of slowing down of the clock, $C^\%$ represents percentage of contraction in the length of the measuring rod, I^l represents increase in the velocity of light, i.e., exceed 186,000 miles per second, D^l represents decrease in the velocity of light, i.e., go below 186,000 miles per second, S^l represents velocity of light, i.e., 186,000 miles per second

Since light particles (photons) do not have mass or inertia, which prevents an object possessing it from accelerating beyond the velocity of light, viz., 186,000 miles per second, theoretically there is nothing to prevent light particles (photons) or other objects without mass or inertia from traveling at a velocity greater than 186,000 miles per second.

The following equation shows that no moving object could travel faster than the velocity of light, which is in accordance with the Special Theory of Relativity:-

$$v = a + b/(1 + ab/c^2)$$

If we let a = velocity of moving train, b = velocity of light beam (which is sent from the back of the moving train to the front of the moving train) with respect to the moving train, which is the moving frame (i.e., the velocity of the light beam (which is sent from the back of the moving train to the front of the moving train) is gauged from the moving train, which is the moving frame), v = velocity of light beam (which is sent from the back of the moving train to the front of the moving train) with respect to the ground level, which is the stationary frame (i.e., the velocity of the light beam (which is sent from the back of the moving train to the front of the moving train) is gauged from the ground level, which is the stationary frame), c = velocity of light = 186,000 miles per second, and, also let $a = b = c$, then:-

$$v = c + c/(1 + c.c/c^2) = 2c/2 = c! \text{ (And not } 2c!)$$

Though theoretically no object could travel faster than the velocity of light because at the velocity of light the object's mass is infinitely great and therefore it is unable to accelerate, an object without mass, possibly, a quantum particle which is somewhat similar to a photon (a photon is a quantum particle without mass always in motion) might be capable of traveling faster than the velocity of light. Such an object or objects might be waiting to be discovered. As it is, a "theoretical" particle which travels faster than the velocity of light, which is termed "tachyon", has been thought to exist.

There have been a number of speculations pertaining to the variable speed of light (VSL), e.g., one theory states that the velocity of light varies with the various stages of the evolution of the universe, exceeding 186,000 miles per second at certain points of time.

Conclusion
The importance of the above points about light cannot be denied, in view of the fact that an important tenet of the Special Theory of Relativity, namely that no object could travel faster than the velocity of light, has just recently been contravened, due to the recent discovery of the superluminal motion of neutrinos at CERN.

6 SPACE-TIME

Time is actually an abstract entity which became part of Special Relativity. There has in fact been postulation that time is unreal but just an invention of the intellect. Whether time is an invention or actually real, it cannot be denied that it plays a very important role in our lives, e.g., without a watch to tell us the time practically all of us would be lost in time. We take a look at the philosophical ramifications and difficulties of time, as well as time-travel, and get things clarified.

Space-Time Continuum
When Einstein first propounded the concept of the space-time continuum (which implies the possibility of time-travel) in general relativity there were critics who scoffed at this revolutionary idea, which posits the existence of four dimensions in the universe: the three physical dimensions of length, breadth and height, plus the fourth dimension of time; they could not conceive time as the fourth physical reality, they could not understand the relativity of time.

What is the actual problem with time? The problem is that while the three dimensions of length, breadth and height are physical and visualizable, the fourth dimension of time is abstract, intangible and not possible to visualize; any object could easily be physically observed or seen moving or traveling along any of the three physical dimensions of length, breadth and height, and, even along all three of these physical dimensions simultaneously (which could be described by a vector diagram, utilizing vector analysis), but it is not possible to physically observe or see an object move or travel in the time dimension, i.e., one could use a ruler to physically measure or gauge the length, breadth and height dimensions but it is not possible to use a ruler to physically gauge the time dimension. However, the time dimension is directly related to the length, breadth and height dimensions - moving or traveling along any or simultaneously all of the three dimensions of length, breadth and height from one point to another point in space requires a certain amount of time which depends on the velocity or speed of movement or travel, this amount of time could only be measured or gauged by a clock or watch. As the three dimensions of length, breadth and height are physical, tangible, visible, concrete, while the time dimension is not physical at all and is strictly intangible, abstract (time is strictly an abstraction), we should not regard or treat the time dimension as similar to the dimensions of length, breadth and height. It is apparent that the confusion caused is due to us treating the non-physical time dimension as similar to the physical dimensions of length, breadth and height, i.e., treating non-physical time as physical (giving rise to the absurd speculation about (physical) time-travel as well). This is actually a philosophical difficulty and should first be tackled. However, movement precedes time, the latter being the measure of the former.

The concept of Einstein's space-time continuum seems to go against common sense. If time is the fourth dimension, can we move or travel in the time dimension as we can do so along the three physical dimensions of length, breadth and height - many appear to think we can? As per the reasons stated just above, this is clearly not possible. Can there be more than four dimensions in the universe? Kaluza, e.g., had modeled a universe with 21 dimensions, reducible to 11 dimensions, a universe which could be described physically in five dimensions. Einstein found his idea elegant and encouraged its publication.

Einstein's concept of time as the fourth dimension of the universe may be "interpreted" in a way which makes more sense (which implies subjectivity) although some people cannot accept it as sensible because time to them is abstract, intangible, and lies outside our physical senses. If a person, e.g., were to enter a completely dark room, whereby he could not see anything at all, and hence, any movement, we can say that he could not see time (passing), and would lose his sense of time. But when there is light, and he could see movement (even the movement of sunlight, moonlight and day and night), e.g., he could see the hands of a clock moving, he could see the points in space, he could see people and objects moving from point to point in space, he could "see" and is in fact "seeing" time (passing), "seeing" here is "sensing" rather than physically seeing, bearing in mind that time is actually intangible and cannot be physically seen or handled as though it were a solid, a physical, a tangible, object. (Here it invokes the question of how time-travel is ever physically possible, though mentally or consciously it is possible - to travel to the past just think of the past and to travel to the future just think of the future.) This gives rise to another philosophical point, viz., the presence of light, which results in visibility and the physically observing or seeing of movements, is necessary for the existence of time (the sense, awareness, consciousness or idea of the passing of time), i.e., light is necessary for sight (we cannot see in the dark), which in turn is necessary for the existence of time (sensation, awareness, consciousness or idea of the passing of time due to the observing or seeing of a physical object moving from point to point or location to location in space). This is an important point to note.

Movement, any movement from point to point in space, e.g., a person's or object's movements from place to place, the earth's orbit round the sun and the orbits of the other planets, is analogous or equivalent to the fourth dimension of time (each of these movements takes a certain amount of time to complete, time which could be measured or gauged only with a clock or watch) - movement is therefore the equivalent of Einstein's fourth dimension (i.e., time) - movement could be translated, or, transformed into time. Hence, Einstein's continuum of the four dimensions of length, breadth, height and time, i.e., his space-time continuum, is equivalent to the length, breadth, height and movement continuum which is visualizable, as movement is tangible, observable, seeable, unlike time.

Time could in a (subjective) way be looked upon as a physical reality and not as a higher reality existing outside our senses, especially our sense of sight, not as such an abstract entity as is deemed, i.e., it is a matter of interpretation. For instance, when we are looking at the physical objects which are making their movements along points in space, we cannot fail to be aware or conscious of time passing. (A moving object is deemed as taking a certain amount of time to move from a certain point to a certain point (distance) in a certain direction, whose actual movement could be observed and gauged in terms of time, distance, velocity, force and direction.) If we touch these objects we could feel the force of their movement, or, on the other hand, the force of our movement (along the objects) and this would also give us a sense of time.

To put it another way, time could be interpreted (implying subjectivity) as seeable by the naked eye and sensable by physical touch (in other words, it is a matter of interpretation). But *time* is actually psychological, not physical - it is the *sense, consciousness,* or, *idea* (of time passing), which is associated with the movements of physical objects in any, or, simultaneously all, of the three physical dimensions (of length, breadth and height, which are also evidently the result of (subjective) interpretation, an idea derived from renowned French mathematician and philosopher Rene Descarte's invention of the co-ordinate system, a system of describing a physical object's location in space by using three co-ordinates or axes, commonly called the *X, Y, Z* axes, which are in fact the *length, breadth, height* axes; the other possible (subjective) interpretation of the dimensions of space is to regard space as a Hilbert space, as is the case with the scientific treatment of the micro-world of the quantum particles - a Hilbert space is a space with an infinite number of dimensions, which would make the interpretation of the dimensions of space more complicated and more intractable). In particular, time is always associated with the movements of the second, minute and hour needles of the clock or watch, which measure or gauge the precise passing of time, without which our sense of time or of time passing would be imprecise and vague.

To circumvent the controversy pertaining to whether time is an invention or actually real, time being actually an abstraction which is evidently derived from the movements of objects that could be physically observed or seen (a physical object is deemed, understood, as taking time to move or travel from one location to another location in space, time which could be gauged by a watch or clock) and which is apparently causing some misunderstanding or lack of comprehension, we could describe space as a space-movement continuum or space-vector continuum (which should be visualizable, more evident, more comprehensible, and less controversial) instead of space-time continuum, which are in fact all equivalent; space may also be described as a space-consciousness continuum instead of space-time continuum as time is directly related to both movement and consciousness, being a psychological entity as is described above, with consciousness being also an intangible, abstract entity like time, but this interpretation would be more abstract, and, more problematic, than the space-time continuum (at least, time can be gauged or measured with a clock or watch, whereas there is no way to gauge or measure consciousness). The velocity or speed of movement of an object is derived by gauging the time the object takes to move or travel from one location to another location in space (distance) in terms of units of measurement for distance such as feet, meters or miles per unit of measurement of time such as second, minute or hour. It is evident here that for time (and distance) to be gauged there has first to be movement. We can actually physically observe or see an object move in the length, breadth, height dimensions, axes or co-ordinates, but we can never physically observe or see the object move in the time dimension (though we can mentally or consciously sense it, the movement of the object from location to location in space is known or understood to take time (an abstract, intangible entity) to do so - time being psychological while the dimensions of length, breadth and height are physical). In any case, no one could say movement is not real but an invention of the intellect, unlike the case of time, as movement is physically observable, seeable and thus indisputable (i.e., visualizable), unlike time. We should also bear in mind that movement can be described by, besides time, distance, velocity (units of distance per unit of time), force (amount of mass per unit of acceleration), and direction, which could all be physically measured or gauged, as is stated above. (In vector analysis, movements of objects with directions are depicted with vector diagrams, a vector being a term which describes an object or force moving or traveling in a certain direction.)

Now, back to the important question: Is time real or is it just an invention of the intellect? Time is like beauty and intelligence; they are all very abstract, intangible, and seem unreal. However, we somehow know they are there when we encounter objects or beings endowed with them. But, unfortunately, we cannot physically count or quantify them as we physically count our coins, and, we, e.g., cannot present a bottle of time to a person, we cannot present a bottle of beauty to him, and, we cannot offer him a bottle of intelligence. How nice if we can do so! On the other hand, we can offer a TV set which physically has length, breadth and height to a bored person to provide him entertainment, we can offer a bottle of soft drink to a thirsty person to quench his thirst, and, we can offer a hungry person a packet of food to satisfy his hunger. The purpose of bringing up these examples, and explaining the philosophical logic above, is to really show that we should not confuse the intangible entities such as time, beauty and intelligence with the tangible entities such as a TV set with length, breadth and height, a bottle of soft drink and a packet of food, and treat them all the same, which would create logical problems. Therefore, to avoid such logical problems, the time dimension, which is abstract and intangible, of the space-time continuum should not be confused with and treated the same as the tangible, physical dimensions of length, breadth and height. Is there some other better way to model space? How about a space-gravity or space-graviton continuum, for example? So, is time real? The intellect can consider it real or consider it an invention of the intellect, i.e., not real. It is a matter of interpretation. Both can be acceptable. All this sounds like Einstein explaining the Theory of Special Relativity, isn't it? Bearing in mind that time, which is intangible, is a sort of consciousness or awareness or conception - it is the consciousness or awareness or idea of the period covered by the distance a moving object moved or traveled (i.e., the consciousness or awareness or idea of the time a moving object took to move or travel from one point to another point in space), if consciousness is real, time is also real. This is yet another interpretation of time, which the author tends to favor.

To really physically travel in the time dimension we have to be able to walk, run, move, travel on time, as we do on the road. Can time be the road for us to physically travel on? All this is absurd. Thus, physical time-travel is absurd. It seems difficult to convince the time-travel die-hards that time-travel is an impossibility. However, it appears that through ambiguities many have been misled into believing that time-travel is possible. We should bear in mind that Einstein himself had thought that time-travel is not possible. More on time-travel below. The author had in the past been confused by all the mumbo-jumbo about time and time-travel and has here gone to some length to explicate the philosophy of time in order to convince the readers, as well as himself, that physical time-travel is absurd and impossible.

Time-Travel
How nice it is to time-travel back to the happy childhood days, the good times, the memorable events! All time-travel enthusiasts are probably aware of the "grandfather" paradox, i.e., travel back in time to the time of one's grandfather and then kill him so that he will not be able to father one's father, or, mother, and, therefore, oneself will not exist. (Such paradoxes may be construed as *reductio ad absurdum* evidence that physical time-travel is an impossibility.)

When Einstein published the Theory of General Relativity ten years after his Theory of Special Relativity, it brought out the possibility of traveling into the past. Taking the mouths of two wormholes, which are tunnels through space made possible by general relativity, and putting one mouth on a space-ship and flying the space-ship off at close to the velocity of light would introduce a time difference between them due to *time dilation, i.e., the clock slowing down, and, the aging process slowing down in the mouth on the space-ship. The person who jumps into the "future" mouth of the wormhole would emerge from the other mouth in the past. (*Time dilation, which is predicted by Einstein's Theory of Special Relativity, states that a person flying off on a space-ship at close to the velocity of light would return home a short number of years later to find that many years had passed back on Earth. In other words, the space-traveler had jumped into the future.)

However, the technical difficulties of implementing the above scheme are considerable, though some physicists have speculated that it may become possible in the centuries to come.

Is the above phenomenon really time-travel? Let us look at an analogous situation involving, e.g., two 60-year-old men, applying the same "aging", time dilation principle; for instance, one of them ages more slowly due to a better genetic make-up, e.g., he still has a head of dark hair and looks 20 years younger, while the other man's hair is all white and he looks his age. Can we say that the younger looking man on meeting the older looking man has jumped into the future? Can we say that the older looking man on meeting the younger looking man has jumped into the past?

Another situation of "time-travel" here. Earth rotates clockwise from east to west, at say, x miles per y second. If John were to travel in the same clockwise direction as Earth's rotation towards the west but at a speed greater than x miles per y second, would he be traveling into the future? If John were to travel in the opposite, anti-clockwise direction towards the east at any speed, would he be traveling into the past?

Are all these time-travels not metaphors, or, illusions? All this is apparently evidence of the inventiveness or creativeness of the contemplating intellect which has been at work (and is apparently subjective).

Conclusion
Hopefully, all the apparent confusion, wooliness and misunderstanding about time and time-travel are straightened out.

7 ROLE OF CONSCIOUSNESS IN PHYSICS

What is the relation between quantum particles, consciousness, the unified field theory and relativity?

According to Dirac, light can be treated as waves or particles. In fact, in quantum mechanics, particles are regarded as waves. The behavior of these particles can be predicted, as it were, and, they are thus known as probability waves or Dirac wave particles. There is a wave/particle duality here. When the particle is not observed (when consciousness is not present), it remains a wave (a probability wave), but upon being observed (when consciousness is present) it becomes a particle.

Hence, the evident importance of the part played by consciousness in quantum mechanics. A number of scientists had postulated that there must be a "cosmic consciousness" pervading the universe; objects spring into existence when measurements are made, measurements which are made by conscious beings, which implies that there must be cosmic consciousness that pervades the universe determining which state we are in - some scientists, e.g., Nobel laureate Eugene Wigner, had argued that this is proof of the existence of God or some cosmic consciousness. Wigner had remarked that it was not possible to formulate the laws of quantum theory in a fully consistent way without reference to consciousness.

Consciousness or "mind-force" is evidently a potent force in nature. The mind is part, an indispensable part, of nature. Scientists such as David Bohm and Werner Heisenberg, as well as many other scientists, evidently understood this fundamental aspect. Classical philosophers such as Berkeley and Hume had wondered whether the existence of any object was independent of the existence of the mind or consciousness: If I had never seen (never been aware of) an object, does that object exist?

Thus, even if a unified field theory or theory of everything were obtained, it will still not give a complete picture of nature if consciousness were excluded. There should therefore be a complementary General Theory of Consciousness. This General Theory of Consciousness will be a very important aspect in our search for the ultimate truth. Many scientists, e.g., David Bohm, Wolfgang Pauli, John von Neumann, Arthur Eddington, Roger Penrose, George Wald, etc., had declared that the universe is mind-stuff. The capabilities of the human mind are so unique that no intelligent machine or artificial intelligence can ever fully duplicate them, according to Sir Roger Penrose, who had authored the books, The Emperor's New Mind, and, Shadows Of The Mind. Could a Supreme Being have created a mind which is capable of questioning its creator, the Supreme Being itself? Will one ever be able to find a computer questioning its creator, the human being?

The unified field theory is only concerned with the following four forces of nature: gravity, strong nuclear force, weak nuclear force and electromagnetism. There may be more than the four forces at work in nature. Some scientists are resurrecting the ether which Einstein had done away with. Can the ether, which is the theoretical medium in which light travels, be a fifth force? Can the centrifugal force of the rotation of the earth be another force affecting nature? Can consciousness, mind power, or psychic energy, be regarded as another force (this evidently applies at the quantum level - recall that according to Heisenberg's Uncertainty Principle the experimenter affects the experiment and vice versa and the experimenter is part of the experiment as well, and, in David Bohm's "looking-glass" universe, the observer is the observed, the part is the whole and different things are really one thing)?

Even Special Relativity is evidently linked to consciousness for it postulates that the intense gravitational field caused by travel at almost the velocity of light will cause the slowing down of clocks and therefore time, as well as the brain (consciousness which feels time passing more slowly) and bodily functions of a person.

It is evident that consciousness is the central player in the scheme of things in nature.

8 GRAVITY

Newton accidentally discovered gravity after observing an apple falling from a tree. What exactly is gravity, which evidently remains a subject of mystery?

Interpretation Of Gravity
Einstein had attempted to unify the four forces of nature, i.e., gravity, weak nuclear force, strong nuclear force and electromagnetism, but had failed. As in the past, he was unable to derive the electromagnetic field equations, even for the weak-field approximation. He was to live to the end of his life without any success with the unified field theory.

To arrive at this grand unified theory a better understanding of gravity would be necessary, e.g., its roles in the micro-world of the quantum particles and in our macro-world. It is thought that quantum particles are free from the effect of gravity, which seems only to have a negligible effect on them, unlike in the macro-world. Nobel Laureate Richard Feynman had posited that gravity could be totally unlike what it had been thought to be, thus possibly making unification with the other three forces impossible. Feynman had suggested that anti-particles are like ordinary particles moving backwards in time, which implies that anti-particles should have anti-gravity. In fact, there is the belief that there is an anti-gravitational force for every gravitational force, just as there is an anti-particle for every particle.

In the Theory of Special Relativity published in 1905, Einstein introduced the gravitational force which was considered responsible for the orbits of planets, as was described by Newtonian gravity. In Newton's law, gravity was an instantaneous force which propagated through space infinitely fast. This was evidently at odds with Einstein's Theory of Special Relativity which postulates that nothing can exceed the velocity of light.

In 1915 Einstein published the General Theory of Relativity which introduced a new theory of gravity that was compatible with the special theory. In this theory the space-time continuum was introduced wherein empty space was likened to a flat rubber sheet which was flexible - a massive object creates an indentation in this empty space, or, "rubber sheet" - this indentation is hence interpreted as a gravitational effect, a curved space-time, a geometry. In fact, in general relativity massive objects affect the way space-time curves. The link between an object's mass and the space-time curvature can be worked out, which is encapsulated in Einstein's all-important "field equations". In this way Einstein was finally able to bring gravity in line with relativity.

Gravitation has always been thought of as a pulling or attracting force, just like the force of attraction between two magnets. Gravitation and magnetism may be different manifestations of the same thing. And, gravity may be a pushing force instead, a force that presses down on all objects in the direction of the centre of the earth. In fact, a push is equivalent to a pull, the former originates at the back of an object while the latter originates at the front. So far, gravitational forces are seen as forces of attraction only, while magnetic and electric forces are forces of attraction and repulsion. There may be a gravitational force of repulsion.

Gravity, which is crucial in the formulation of a unified field theory, can be described by the following formula:-

$$F = G((M_1 \ M_2)/R^2)$$

where F is the gravitational force of attraction, G is the gravitational constant, M_1 and M_2 are the masses of two objects and R^2 is the distance between masses.

Before Newton discovered gravity, nobody had known it existed or had thought that there was such an attractive force. However, Einstein in his General Theory of Relativity interpreted it as a curvature of the space-time continuum, a geometrical form, as is described above. Can gravity be the fifth dimension, in addition to the four dimensions of General Relativity comprising of the three physical dimensions and the time dimension, as Theodor Kaluza had suggested, which impressed Einstein greatly?

According to Einstein's theory of gravity, the hypothetical quantum of gravity, the graviton, which is a spin-2 boson, interacts extremely weakly with other matter, far more weakly than neutrinos; it is so weak that no instruments so far have been able to detect it. In the supergravity extension of this theory of gravity, the graviton finds a superpartner, the gravitino, which is a spin-3/2 fermion. Under local supersymmetric transformations these two particles transform one into the other. When quantum calculations were carried out using supergravity theory, it was discovered that the infinities which plagued the earlier gravity theory with only the graviton were now being cancelled by equal and opposite infinities produced by the gravitino. This is evidently the result of the deeper consequence of the presence of supersymmetry. Though it is not certain whether the supergravity theory is completely renormalizable, this "softening of the infinities" appears to be a step toward a viable theory of quantum gravity. As simple supergravity theory includes only the graviton and the gravitino, this hardly corresponds to the real world with its many particles. Most of those who have worked on supergravity feel that some crucial idea is still missing. Without this crucial idea the theories simply do not describe the real world.

How do we make supergravity theory realistic? If we can solve this problem, we can have supergravity theory as a completely unified field theory. It has been shown that the principle of local supersymmetry is so restrictive that only eight possible supergravity theories exist, which are each labeled by an integer $N = 1, 2$ . . . 8. Supergravity theory shares the same features with its progenitor, the Theory of General Relativity, namely, conceptual power and mathematical complexity. Perhaps, by postulating the existence of a single master supersymmetry we can have a unified field theory that accounts for the whole universe.

The following is Einstein's equation for General Relativity:-

$$G_{im} = -K(T_{im} - 1/2g_{im}T)$$

This beautiful equation expresses the curvature of space-time. The left-hand side refers to a set of terms which characterize the geometry of space, while the right-hand side refers to a set of terms which describe the distribution of energy and momentum, i.e., the left is the geometry side, while the right is the matter side. Reading from left to right is space-time telling mass how to move, while reading from right to left is mass telling space-time how to curve. In General Relativity, there is neither absolute time nor space and gravitation is not a force, or, pull between one object and another but a property of space and time. All this represents a great conceptual leap by the theory's creator, Einstein. As for the coordinate system of Einstein's General Theory of Relativity it has no basis in reality and is only a mental construct used to describe the space-time continuum of the General Theory of Relativity.

One may wonder when a unified field equation will be discovered. However, without a good understanding of gravity, this unified field equation will not come by easily. Once the experimental confirmation of the existence of the graviton, the hypothetical quantum of gravity, is achieved, we should be surer of obtaining a unified field equation that accounts for the whole universe, which may be supported by the existence of a single master supersymmetry. Superstring Theory now appears to pave the way towards achieving this difficult goal.

The search for gravitational waves is still ongoing. The existence of gravitational waves so far is inconclusive - there is only indirect evidence of their existence. In the gravitational effect known as frame dragging, objects occupying the space near to a rotating object get swept around with it; rotating objects drag space around with them rather like a spoon in treacle (syrup). In 2004, scientists analyzing data from two Earth-orbiting satellites apparently found evidence of frame dragging - they claimed to have detected the minute frame dragging effect of our planet. There is the possibility of extracting useful energy from the rotation of our planet. There is speculation that this could serve as a power source for an advanced civilization.

Is Gravity Really What It Is Thought To Be?

Is all the space occupied by us and around us really one immensely large, flexible sheet of curved space-time as is posited by the General Theory of Relativity? Or, is there an infinitude of such "flexible sheets"? That is, is gravity just one flexible sheet of curved space-time or more, perhaps infinitely more? Earth is a sphere but all of us are evidently "stuck to this sphere" and do not fall off it. What is the force, if it exists, which causes this? Will this force be an attractive or pulling force, or, a pushing force? Can gravity be a kind of fluid, whether air, gas or liquid? (We can descend, glide or fly up through the air like how a bird or an aeroplane does. We can dive into and descend in, surface in, float in, and swim in, a pool of liquid. All these may be interpreted as manifestations of "gravitational effect".) Can gravity be electrical, electronic, quantum or nuclear? Or, is gravity none of these but something else, as Richard Feynman had speculated?

It could be any intelligent person's guess.

9 MORE ON GRAVITY

What exactly is gravity? Any theory about gravity would have to be confirmed by physical experiments of course.

Einstein had suggested that gravity is the geodesic of an "imaginary rubber sheet" which occupies space in his paper on general relativity. By this suggestion he had transformed gravity, which is an abstract entity, into a geometric, more tangible entity that is linked to what he described as space-time. This had been evidently an attractive idea to his peers. How should this "rubber sheet" be visualized or interpreted? Is this "rubber sheet" really three-dimensional as has been illustrated by pictures in scientific tomes, or, is it of more dimensions, say having infinite dimensions?

The author would like to modify this "rubber sheet" idea of Einstein's. The suggestion here is to replace this "rubber sheet" with a layer (or layers) of fluid, which is possibly a combination or compound of air, liquid/vapors, gases, chemical elements, particles, etc., which has the quality of adhesiveness, i.e., it could adhere to matter or any object it comes into contact with (like water drenching all objects it comes into contact with), perhaps including light particles and quantum particles. This "sheet which replaces Einstein's geometric rubber sheet" is comparable to a "sea of water or parcel of air" in which matter occupies. The activity of any matter within this "sheet" could be compared to that of a submarine, fish or swimmer navigating in the sea of water or bird, airplane or glider navigating in the parcel of air. As is stated above, this "sheet" or layer could possibly adhere to all matters it comes into contact with, surrounding and enveloping (drenching) these matters. Since its shape would depend on the shapes of the objects (matters) it surrounds and envelops which could be all kinds of shapes the shape of this "sheet" or layer should be fluid, flexible and infinite in dimensions. Denser objects could congregate to the bottom of this "sheet" or layer while less dense objects could float nearer or at the top (like denser objects sinking to the sea-bottom while less dense objects float nearer or at the top of the ocean). All of these could be interpreted as "effects of gravity".

We now look at the case of outer space and other planets. Like Einstein's "geometric rubber sheet of space" this "sheet" or layer could stretch all the way from earth to outer space, surrounding and enveloping (drenching) the planets out there (like water twirling around in outer space and gushing all over the planets). The "sheet" or layer could be interpreted as denser in outer space allowing space travelers to float in it, which could all be interpreted as the "lesser effect of gravity".

The above is the germ of an idea about the real nature of gravity.

10 THE PUZZLE OF QUANTUM ENTANGLEMENT

The phenomenon of quantum entanglement involving two particles has been a mystery for a long time. How can quantum entanglement be explained?

Entangled Particles
Quantum entanglement is a phenomenon wherein the quantum properties of two particles become codependent, with the properties of one being instantaneously affected by measurements conducted on the other.

Entangled systems need special preparation, e.g., a pair of electrons having opposite spins, as is specified by the Pauli exclusion principle, has to be created, with the actual spin of each particle remaining in a state of quantum uncertainty (a situation described as "entanglement of the wavefunction" by Erwin Schrodinger) On the separation of the pair of particles, even by a huge distance, and on measuring one particle's spin the other particle's spin will automatically resolve itself in the other direction. This effect occurs instantaneously, apparently breaching the velocity of light and the rules of relativity, a phenomenon that Einstein referred to as "spooky action at a distance".

In quantum teleportation, a pair of entangled particles are used to transmit information about a third object instantaneously from one place to another. The original particle with information to be teleported is scanned. The scanning process disrupts the original particle and modifies both the entangled particles, which are separated by a large distance, instantaneously. The treatment process recreates the properties of the original particle. A "teleported" replica is thus formed.

Possible Explanations
There are several possible interpretations or explanations for the behavior of entangled particles, which are as follows:-

First Interpretation
The two entangled particles may be linked by some kind of electromagnetic "force/link", the analog of which is the mechanical system of two similar physical objects, e.g., two similar balls or two similar wheels, linked by a rod. For instance, one of these two similar objects, e.g., two similar balls, is directly joint, connected, to one end of the rod while the other object (ball) is joint to the other end of the rod through two similar interlocking gears which are mechanically arranged in such a way that the turning of one of these objects (balls) at one end of the rod by a certain fraction of a revolution in one direction would result in the object (ball) at the other end of the rod turning by the same fraction of a revolution in

the opposite direction at the same instant (i.e., instantaneously). What happens is that turning, e.g., the first object (ball) joint or connected directly to (one end of) the rod would turn the rod in the same direction by the same fraction of a revolution at the same instant, the rod would turn the first gear joint to it at its other end in the same direction by the same fraction of a revolution at the same instant, this gear would turn the similar gear interlocked with it in the opposite direction by the same fraction of a revolution at the same instant, and, as the other object (ball) is joint to this second gear that turns in the opposite direction the other object (ball) itself also turns in the opposite direction by the same fraction of a revolution at the same instant (all these various actions taking place at the same time, all at once, simultaneously). On the other hand, if the second object (ball) joint to the second interlocked gear were turned instead it would cause the first object (ball) joint directly to the rod at the other end to turn in the opposite direction by the same fraction of a revolution at the same instant. This analogous mechanical principle might apply to the behavior of the two entangled particles, whose "spooky action at a distance" is however abstract and invisible to the eyes unlike the above-described mechanical action which is visible to the eyes.

The following describes how the above-stated mechanical principle might apply to the behavior of the two entangled particles. Any spin motion (measured) in one of the particles may theoretically cause instantaneous motion (e.g., spin or vibratory) in the electromagnetic "force/link" that links this particle to the other particle (as per the case of the first object (ball) and the rod in the above-described mechanical example). This instantaneous motion of the electromagnetic "force/link" may theoretically effect instantaneous motion in the other particle (as per the case of the rod and the second object (ball) in the above-described mechanical example) which may spin in the opposite direction (as it has been conditioned to do so through the entanglement process in accordance with the Pauli exclusion principle). (Note Carefully: The motions of the two entangled particles and the electromagnetic "force/link" may theoretically take place simultaneously, at the same instant or instantaneously (as is in the case of the moving objects/parts in the above-described mechanical example)).

Second Interpretation
The two entangled particles may theoretically be simultaneously controlled by a "brain" or "controller". This "brain/controller" may theoretically issue a signal to both particles at the same instant causing them to act as they do at the same instant. This is comparable, e.g., to a computer issuing a command to two printers (or other equipment) at the same instant causing the two printers to print at the same instant (parallel processing comes to mind), with the two printers programmed to respond differently to the same command at the same instant (e.g., one printer prints blue ink in response to a command while the other printer prints red ink in response to the same command at the same instant).

Third Interpretation

Information from one of the two entangled particles may theoretically be carried to the other particle by an extremely fast carrier wave that travels faster than the velocity of light causing the other particle to act with an opposite spin at practically the same instant. (Note: Since the speed of this carrier wave theoretically exceeds the velocity of light and light may be required to detect it, it may be undetectable.)

Fourth Interpretation

There may theoretically be an unknown influence, a mysterious undiscovered force, at work.

Programming

Equipment controlled by computers such as robots have to be programmed to get them to work in a certain way, e.g., one programming method involves walking the robot through the operating sequence to "teach" it the operating sequence. Entanglement of two particles is rather similar to the programming of an equipment with a computer resulting in the two particles acting the way they are expected or "programmed" to.

11 UNIFIED FIELD THEORY

How may a unified field theory and theory of everything be obtained?

Introduction

Having only a beautiful theory and beautiful equations to model nature are insufficient for affirming the characteristic of the universe. The affirmation has to be ultimately carried out by physical experiment. However, in lieu of physical experimentation, computer simulation is proposed, simulation with powerful software having been effectively utilized in areas such as aeronautical, electronic, mechanical and marine designs, wherein it now becomes unnecessary to produce costly prototypes. A few possible ways of unification are presented hereafter.

Forces Of Nature And Unification

Our views of nature may be modified in the future. For instance, superstring theories, which had been neglected in the past, are now the "in" thing, being regarded by many scientists as beautiful, or, elegant, and a possible theory of everything.

Einstein had attempted to unify the four forces of nature, i.e., gravity, weak nuclear force, strong nuclear force and electromagnetism, but had failed. As in the past, he was unable to derive the electromagnetic field equations, even for the weak-field approximation. He was to live to the end of his life without any success with the unified field theory.

Einstein had thought that David Bohm would be the first scientist to solve the unification problem. But ironically the latter regarded the concern with the unified field theory problem as merely an unnecessary fuss, dismissing it as an "illusion of parts" and simply relegating the problem to the logical constraints of topology; he resorted to the metaphysical way of interpreting the universe, calling it the "looking-glass" universe.

However, gravitation and electromagnetism can be linked and the result can be anti-gravitational force and/or torsion in space-time. There is also the belief that there is an anti-gravitational force for every gravitational force, just as there is an anti-particle for every particle. Nobel Laureate Richard Feynman had suggested that anti-particles are like ordinary particles moving backwards in time, which implies that anti-particles should have anti-gravity.

Gravitation has always been thought of as a pulling or attracting force, just like the force of attraction between two magnets. Gravitation and magnetism may be different manifestations of the same thing. And, gravity may be a pushing force instead, a force that presses down on all objects in the direction of the centre of the earth. In fact, a push is equivalent to a pull, the former originates at the back of an object while the latter originates at the front. So far, gravitational forces are seen as forces of attraction only, while magnetic and electric forces are forces of attraction and repulsion. There may be a gravitational force of repulsion. All this will affect our approach towards the unified field theory.

There is evidently a "looking-glass" characteristic in the universe. David Bohm, the British scientist whom Einstein had once hailed as being the person who would one day solve the unified field theory problem, had postulated that our universe is a vast fluid nothingness or no-thingness in which everything is. He had spawned a surprising relationship between maps and terrains. According to him, in the "looking-glass" universe, our mapmaking changes the very terrain, and the terrain in turn changes our map; maps, mapmakers and terrains intertwine to form an integral whole. In other words, different things are really one thing. Heisenberg's Uncertainty Principle postulates that the experimenter affects the experiment and vice versa, and the experimenter is also part of the experiment. Accordingly, all things are relative, how things are really depends on how or from what angles we look at them; there is a "yin" to every "yang", and the "yin" and the "yang" combine to make the "yin-yang" which is an integral whole - so good cannot be distinguished from bad for they belong to the same whole, the "yin-yang". Regarding Einstein's unified field theory, Bohm maintained that it is based on the illusion of parts (gravity, strong nuclear force, weak nuclear force and electromagnetic force) and that it is a futile problem. The observer is the observed, the part is the whole - all this seems more metaphysics than physics, but nevertheless this has evidently been the state of physics. Bohm believed that unification could be expressed by using the logical relations of topology, as is stated above.

Failure
The failure in deriving the unified field theory is the failure to derive an equation which will link our visible macro-world with the invisible micro-world of the quantum particles, which will link the gravitational force with the weak nuclear force, strong nuclear force and electromagnetism, an equation which should encapsulate the totality of information about the universe. If Bohm's interpretation were correct, all these four forces are actually one and the same force, being different manifestations of the same force. Can any of these four forces exist without the others? They are all evidently essential parts of our universe, making up the whole.

Gravity
Gravity, which is crucial in the formulation of a unified field theory, can be described by the following formula:-

$$F = G((M_1 \ M_2)/R^2)$$

where F is the gravitational force of attraction, G is the gravitational constant, M_1 and M_2 are the masses of two objects and R^2 is the distance between masses.

Before Newton discovered gravity, nobody had known it existed or had thought that there was such an attractive force. Einstein in his General Theory of Relativity interpreted it as a curvature of the space-time continuum, a geometrical form. Can gravity be the fifth dimension, in addition to the four dimensions of General Relativity comprising of the three physical dimensions and the time dimension, as Theodor Kaluza had suggested, which impressed Einstein greatly?

Hilbert Space

Can the physical world be considered infinite-dimensional, i.e., a Hilbert space, instead of four-dimensional, a commonly held view? It depends on how we look upon the physical world, on our mental inventiveness, and is thus subjective. Quantum particles in the micro-world, unlike the objects in the macro-world, are comparatively unpredictable where their actions or movements are concerned and can only be predicted if at all in a probabilistic fashion. We will never be able to know for certain where a quantum particle will turn up next. Moreover quantum particles are capable of "entanglement", and, also teleportation, which is "spooky" and incomprehensible. We can have a quantum field equation involving infinite dimensions. According to modern quantum mechanics, all possible physical states of a system correspond to space vectors in a Hilbert space. An infinite-dimensional Hilbert space will also fit in with the theory of the existence of an infinite number of parallel universes which are connected with each other through worm-holes.

Supersymmetry

According to Einstein's theory of gravity, the hypothetical quantum of gravity, the graviton, which is a spin-2 boson, interacts extremely weakly with other matter, far more weakly than neutrinos; it is so weak that no instruments so far have been able to detect it. In the supergravity extension of this theory of gravity, the graviton finds a superpartner, the gravitino, which is a spin-3/2 fermion. Under local supersymmetric transformations these two particles transform one into the other. When quantum calculations were carried out using supergravity theory, it was discovered that the infinities which plagued the earlier gravity theory with only the graviton were now being cancelled by equal and opposite infinities produced by the gravitino. This is evidently the result of the deeper consequence of the presence of supersymmetry. Though it is not certain whether the supergravity theory is completely renormalizable, this "softening of the infinities" appears to be a step toward a viable theory of quantum gravity. As simple supergravity theory includes only the graviton and the gravitino, this hardly corresponds to the real world with its many particles. Most of those who have worked on supergravity feel that some crucial idea is still missing. Without this crucial idea the theories simply do not describe the real world.

How do we make supergravity theory realistic? If we can solve this problem, we can have supergravity theory as a completely unified field theory. It has been shown that the principle of local supersymmetry is so restrictive that only eight possible supergravity theories exist, which are each labeled by an integer $N = 1, 2 \ldots 8$. Supergravity theory shares the same features with its progenitor, the Theory of General Relativity, namely, conceptual power and mathematical complexity. Perhaps, by postulating the existence of a single master supersymmetry we can have a unified field theory that accounts for the whole universe.

General Relativity And Curved Space-Time
The following is Einstein's equation for General Relativity:-

$$G_{im} = -\mathbf{K}(T_{im} - 1/2g_{im}T)$$

This beautiful equation expresses the curvature of space-time. The left-hand side refers to a set of terms which characterize the geometry of space, while the right-hand side refers to a set of terms which describe the distribution of energy and momentum, i.e., the left is the geometry side, while the right is the matter side. Reading from left to right is space-time telling mass how to move, while reading from right to left is mass telling space-time how to curve. In General Relativity, there is neither absolute time nor space and gravitation is not a force, or, pull between one object and another but a property of space and time. All this represents a great conceptual leap by the theory's creator, Einstein. As for the coordinate system of Einstein's General Theory of Relativity it has no basis in reality and is only a mental construct used to describe the space-time continuum of the General Theory of Relativity.

A suggestion is to change the left-hand side of the equation, the geometry of space, which is here a four-dimensional space-time continuum, into an infinite-dimensional space, a Hilbert space, which is as follows:-

$$G_{imH} = -\mathbf{K}(T_{im} - 1/2g_{im}T)$$

What kind of geometrical form can represent this infinite-dimensional space? One geometrical form of this nature can be an infinite number of Moebius Strips which are intricately intertwined and linked with each other (with each Strip being cut lengthwise into several narrower strips that are connected together at narrow points, which represent parallel universes). It is thought that since we are only able to move around in the three large, observable spatial dimensions comprising of length, breadth and height, and one of time, all other dimensions must be very small and thus invisible to us, being curled up in a multidimensional space (which may be construed as representing the invisible micro-world of the quantum particles). This is in keeping with the concept of the unified field theory which Einstein had attempted to formulate by combining General Relativity and quantum theory.

Success

One may wonder when a unified field equation will be discovered. However, without a good understanding of gravity, this unified field equation will not come by easily. This unified field equation will of course link both the macro-world and the micro-world (through gravity, which will be the common denominator for both). Once the experimental confirmation of the existence of the graviton, the hypothetical quantum of gravity, is achieved, we should be surer of obtaining a unified field equation that accounts for the whole universe, which may be supported by the existence of a single master supersymmetry. Superstring Theory now appears to pave the way towards achieving this difficult goal.

Universal Laws

Can there be some yet to be discovered universal laws which govern everything that exists in the universe? It is difficult to tell but it will be very useful to know these laws. This then will really be the theory of everything.

To many, including scientists, even Einstein, the laws of nature had been created by a Supreme Being, a God (whom Einstein believed does not play with dice). Alternatively, can all of nature be a computer simulation carried out by a very advanced race of beings? In 2001, a philosopher called Nick Bostrom had begun circulating a paper titled "Are You Living In A Computer Simulation?" This has been considered a good possibility.

Practical Affirmation

It appears a good idea to get some computer game designers and/or computer programmers to collaborate with the scientists to produce a simulation of life and existence in the universe. This is probably a mammoth if not impossible task. These programming experts and scientists have to sort of play God. What will be the parameters involved, the coordinate system to be utilized (e.g., three-dimensional, three-and-a-half dimensional, four-dimensional, etc.), the algorithms or mathematical formulas to be used for governing the movements of virtual objects (which will be important as they may be equivalent to the field equations of the unified field theory), and so on? With such a simulation, we may be able to understand better how the forces of nature, e.g., the mysterious but all-important gravitational force, behave. In this manner, which can be considered a kind of reverse engineering, success with the unified field theory and the theory of everything may be achieved.

Conclusion

The missing link in this unification process is gravity, which is still not a fully understood phenomenon. We like to think that gravity affects our position and movement in space. But gravity may not be what we think it is. It may not even exist, which means that our search for gravitational waves and gravitons may be a futile exercise. The search for gravitational waves is still ongoing. The existence of gravitational waves so far is inconclusive - there is only indirect evidence of their existence. In the gravitational effect known as frame dragging, objects occupying the space near to a rotating object get swept around with it; rotating objects drag space around with them rather like a spoon in treacle (syrup). In 2004, scientists analyzing data from two Earth-orbiting satellites apparently found evidence of frame dragging - they claimed to have detected the minute frame dragging effect of our planet. There is the possibility of extracting useful energy from the rotation of our planet. There is speculation that this could serve as a power source for an advanced civilization.

Is all the space occupied by us and around us really one immensely large, flexible sheet of curved space-time as is posited by the General Theory of Relativity? Or, is there an infinitude of such "flexible sheets"? That is, is gravity just one flexible sheet of curved space-time or more, perhaps infinitely more? Earth is a sphere but all of us are evidently "stuck to this sphere" and do not fall off it. What is the force, if it exists, which causes this? Will this force be an attractive or pulling force, or, a pushing force? Can gravity be a kind of fluid, whether air, gas or liquid? (We can descend, glide or fly up through the air like how a bird or an aeroplane does. We can dive into and descend in, surface in, float in, and swim in, a pool of liquid. All these may be interpreted as manifestations of "gravitational effect".) Can gravity be electrical, electronic, quantum or nuclear? Or, is gravity none of these but something else, as Richard Feynman had speculated?

What we regard or interpret as the attractive force known as gravity may be just a facet of electromagnetism, which comprises of an electric field and a magnetic field, wherein one mass attracts or repulses another mass and pulls or pushes the latter towards or away from itself (a pull/attraction and a push/repulsion may be viewed as equivalent, the former being a force exerted in front of a mass to move it while the latter is a force exerted behind the same mass to move it in the same direction). If this were the case, the weak nuclear force, strong nuclear force and electromagnetism would be sufficient to account for the phenomena of both our macro-world, the universe, and the micro-world of the quantum particles. The gravitational wave and the quantum of gravity, the graviton, may be non-existent and may never be found, though evidence of their existence is still being sought after; a good reason why they are not directly found (the gravitational wave is only indirectly deduced) can be that they are non-existent, and, when physical experiments fail to detect them directly (though it is thought that they are very weak and instruments are not sensitive enough to detect them) after a long period of time we should be prepared to make a turnaround instead of continuing banging the head against the wall.

There is another possibility. All space wherein both the macro-world of the universe and the micro-world of the quantum particles occupy may be a sea of invisible, yet undetected, material or fabric, much as all space is occupied by fluid such as air, liquid or gas. This material may be called "ether" or any other terms.

This is the substance where all matter in the macro-world and quantum particles in the micro-world are theoretically submerged in (this submerging effect being manifested possibly as the gravitational effect evident in the universe), much as objects exist and perambulate in fluids such as air, liquid or gas. This is theoretically the most basic constituent of space such that a vacuum in space is theoretically non-existent. The ether, which has been contemplated heretofore, is apparently comparable to this material or fabric, which may be regarded as the fabric of the cosmos. This will theoretically become the link between all the forces of nature in both the macro-world and the micro-world, without which nothing can exist, similar to the case of the very basic life-giving oxygen without whose existence in the air or in liquid no animals, insects or marine life can exist. This will probably be the spark-of-life, the life-giving, the activity-causing, the mobility-causing, the motion-causing, the behavior-causing, the action-causing, agent which causes activity, motion, change in all particles and matter, e.g., compositional and positional/directional/gravitational changes, including possibly consciousness, which may be regarded as the essence, soul or spirit of existence or life. Quantum particles in the micro-world apparently have a "life" of their own, e.g., light particles or photons are always in motion, like the animate objects in the macro-world. What is the cause of this apparent life in the macro-world of the universe and the micro-world of the quantum particles if not for this life-giving, action-causing essence which for want of a better term may be called "ether", "life-giver", "life-force", "life-fluid", "change agent", "activator", "essence", "energizer" or any other suitable term? The cells which make up our bodies, the bodies of animals and the bodies of plants are apparently living ones using up oxygen to generate energy while quantum particles also generate energy. This possible agent, which we may, e.g., call "life-fluid", which probably causes all this activity in the cells and the quantum particles can be regarded as the common link between these two things found respectively in the macro-world and the micro-world. The forces in both the macro-world and the micro-world may thus be unified in this manner. However, detecting and confirming the existence of this theoretical life-giving, action-causing agent common to both the macro-world and the micro-world is likely to be a hurdle. We can now only define it abstractly as an agent which causes activity, motion, change in particles and matter including possibly consciousness. This theoretical agent may also be regarded as the controller of nature, the director which guides all activities in nature, like how the DNA regulates bodily functions and the computer software regulates computer functions. Figuring out what will constitute this important agent, e.g., whether it is a form of energy, whether it is made up of some particular atoms, whether it is a sort of fluid, and so on, is a difficulty, which has to be resolved before experiments can be carried out to detect it.

12 THEORY OF EVERYTHING (TOE)

A theory of everything , or, grand unified theory (which Einstein had been working on without success, with Superstring Theory now being a good candidate), is one which unites all the forces of nature, viz., gravity, electromagnetism, the strong nuclear force and the weak nuclear force. Important as this theory might be, it is lacking in one important fundamental aspect, viz., the role of consciousness, which could in fact be considered the most fundamental aspect of physics. This chapter explains that a theory of consciousness is more important than a theory of everything or grand unified theory and should be the theory of everything instead, or, at least, a part of the theory of everything.

An Alternative Theory Of Everything?

It is apparent that consciousness is somehow connected with phenomena in the quantum realm. Consider Bell's double-slit experiment. In this experiment, are the particles and waves just sensitive to screens and other equipment or are they picking up messages from the physicist's brain? It could be said that particles of matter and particles of mind might come into being together, but in any case our self-awareness and what seems to be some kind of consciousness at the quantum level appear to be in deep communication. David Bohm had in his classic work, Wholeness And The Implicate Order, developed a theory of quantum physics which treats the totality of existence, including matter and consciousness, as an unbroken whole. There is the implication that at the sub-quantum level the observing device used to measure the quantum particles must have connections with all parts of the system, including the link with our consciousness, and through these a "signal" might be transmitted to the molecule that a certain observable was eventually going to be measured. Consciousness is non-local; in other words, no one could say where the mind is or how far the effects of thought could reach. As a matter of fact, within the brain, all the forces are active, including gravity, which is the force that holds the entire universe together.

Though fundamentally the material brain and the other matters are comprised of the same thing, viz., atoms, reductionists might wonder why the material brain as compared to other matters is so special. Is this due to the special composition of the atoms in the brain? If we take a number of atoms and arrange their composition so that they would be similar to that of the material brain, could we produce a brain, and, consciousness?

The universe might be in some sense a Great Mind and a theory of everything might have to include a theory of consciousness. Superstrings, which are a strong candidate for a theory of everything, might be thought particles with a life of their own. Many physicists are making attempts at deriving a Grand Unified Theory of the universe on the basis of particle physics. This effort might be incomplete as particles might be just a reflection of the information-processing foundations of the universe (but it is certainly not a waste of time as this research might help us to figure out how the information-processing system works). In the last analysis, we might not be able to completely understand the universe, if it is ever possible to do so, until it is examined as a self-evolving and organising information-processing machine, one which produces intelligent minds to examine itself with. Hence, a theory of consciousness might be consolidated with the theory of physics (such as the Superstring Theory or the Membrane Theory) into a Grand Information Theory (GIT). This could be considered the Theory of Everything.

In quantum mechanics, the behavior of particles, which are regarded as waves, could be predicted, as it were, and, they are thus known as probability waves or Dirac wave particles. Here, there is a wave/particle duality. When the particle is not observed (when consciousness is not present), it remains a wave (a probability wave), but on being observed (when consciousness is present) it becomes a particle. (Note that according to the Uncertainty Principle, for which Werner Heisenberg won a Nobel prize, the very act of observing a quantum particle affects its behavior, i.e., consciousness affects a quantum particle.) Pauli, who was a Nobelist, was fascinated by subatomic particles and consciousness, collaborating for some time with psychologist Carl Jung, whose patient he was for a time. The mathematician, John von Neumann, the biologist, George Wald, and the physicists, David Bohm and Arthur Eddington, had declared that the universe is mind-stuff. The mathematical physicist, Sir Roger Penrose, sometime colleague of Stephen Hawking, considers that there is "definite possibility" that consciousness is connected with phenomena in the quantum realm. The Anthropic Principles (both strong and weak) stipulate that in man there are intellectual capacities which are there for a reason, that somehow human beings with their minds are obliged to help the universe through the next stage; this is indeed manifested by the fact that scientists such as Einstein, etc., had been using their brains/consciousness to understand nature and many had been attempting to formulate a theory of everything or unified field theory. The Gaia Hypothesis of biologist, James Lovelock, paints a picture of the earth functioning as one large organism, which implies will and consciousness being at work.

A number of scientists had postulated that there has to be a "cosmic consciousness" pervading the universe; objects spring into existence when measurements are made, measurements which are made by conscious beings, which implies that there must be cosmic consciousness that pervades the universe determining which state we are in. Some scientists, e.g., Nobel laureate Eugene Wigner, had argued that this is evidence of the existence of God or some cosmic consciousness. Wigner had remarked that it was not possible to formulate the laws of quantum theory in a fully consistent way without reference to consciousness.

Classical philosophers such as Berkeley and Hume had in fact questioned whether the existence of any object was independent of the existence of the mind or consciousness: If I had never seen (never been aware of) an object, does that object exist?

Thus the great importance of the role of consciousness in nature. In fact, there appears to be an intricate link between nature and consciousness, the latter being apparently the common denominator in the workings of nature, especially at the quantum level, as is described above. However, far-fetched as it might seem, consciousness could be some sort of particles, not unlike quantum particles, which could explain why they are able to interact with one another, as is afore-described. This could also possibly explain phenomena such as intuition, mind-reading and telepathy. For example, Rupert Sheldrake, a well-known biologist, in his "morphogenesis" theory, stated that all our minds or consciousness are linked or interconnected, so that how people think and behave in one geographical area affects how people living in another distant geographical area act and think without any communication whatsoever between them, a phenomenon which applies to animals as well. Consciousness could therefore be regarded as a "force" of nature.

We present an important poser here: What is life, the theory of everything or the grand unified theory without consciousness? This is in fact an irrelevant question as it is a tautology, for life and consciousness are synonymous, and, without life or consciousness to contemplate a theory of everything or grand unified theory, the latter is an impossibility and is redundant.

It is foremost, most important, to have consciousness, and, hence a theory of consciousness, which should precede a theory of everything or grand unified theory, for to contemplate a theory of everything or grand unified theory without taking into account the evident role of consciousness is like riding the horse-cart without the horse. Isn't it more important to better understand ourselves, our consciousness, our nature, first, an understanding we are still evidently lacking, before we try to probe further the nature external to us? Isn't it evident that it is mind which controls external matter and not vice versa, for otherwise life would be overwhelmed by natural phenomena? Therefore, shouldn't a theory of consciousness really be the theory of everything, e.g., as a Grand Information Theory (GIT), which is described above, or, at least, a part of the theory of everything, a very important, fundamental part, consciousness and matter having some fundamental link as is described above? Importantly, this theory of consciousness giving us a more comprehensive understanding of our mind would be a boon or aid to our affairs.

13 MORE ON THE THEORY OF EVERYTHING (TOE)

The conventional theory of everything concerns the unification of the four forces of nature: gravity, weak nuclear force, strong nuclear force and electromagnetism. But some of the most fundamental aspects of nature are left out. This chapter raises some important points which ought to be covered by a new theory of everything.

Firstly, let us take a look at infinity which is an important idea in mathematics. Is outer space really infinite? If on the contrary outer space were finite and there is a limit or boundary to how far one could travel there, what structure/support (space) would this whole finite space be positioned on. Next, what structure/support (space) would the structure/support (space) of this whole finite space be positioned on, and, what structure/support (space) would the structure/support (space) of the structure/support (space) of this whole finite space be positioned on, and so on and on to infinity? In this case there would be an infinitude of structures/supports (infinitude of spaces on spaces). [Compare: Zeno's paradox] In either case, infinity comes into the picture. The infinity of outer space thus makes sense. Is infinity, e.g., the infinity of outer space, indeed a real physical phenomenon and not a mere concept? The physical or actual existence of infinity implies the possibility of infinite powers such as those deemed to be possessed by a Supreme Creator. The problem here is that it is impossible to physically confirm infinity even if one has an infinite life-span simply because infinity has no end and the search for infinity would be endless, like a cat chasing its own tail.

Secondly, is consciousness indeed a driving-force behind nature? It appears that it is unscientific or unconventional to consider consciousness as a driving-force behind nature and it is best ignored. But in quantum theory the behavior of quantum particles could only be described in a probabilistic manner with no certainty, probability being a function of consciousness; Heisenberg's Uncertainty Principle postulates that the viewer (i.e., consciousness) affects the viewed quantum particle and vice versa. How to deny the importance of consciousness in science and shouldn't there be a comprehensive theory of consciousness? Without life or consciousness there would not be the need, and, ability, to do science and control nature. A consciousness probably much different from and much more superior than human consciousness, e.g., aliens or extra-terrestrials, should not be discounted.

Finally, are all life-forms really the result of evolution, or, are they the handiwork of a Creator. To many, life-forms are so complex that they could not possibly have just come into existence spontaneously without the intervention of a Creator, though computer simulation has shown that evolution is possible. How to account for the existence of the Creator? By the same reasoning given just above could He be the handiwork of a Higher Creator, who in turn is the handiwork of an even Higher Creator, and so on and on to infinity. The ultimate answer could be that everything which exists is the handiwork of an infinitely powerful Supreme Creator, who is the First Cause. Those who believe in God would abide by this theory. However there are a number of atheists who do not subscribe to this and many more who neither believe nor disbelieve this. A computer or robot which has been created by man does not question its maker unlike the atheists. It is possible that a Supreme Creator has also created atheists. The idea often put forward is that man has been given Free Will by his Maker to do whatever he likes. How to scientifically explain all this one way or the other? Going further, how did the DNA or "computer program" which is responsible for the development and characteristic of living things arise; who created these "computer programs"? Are there "DNAs" or "computer programs" governing the behavior of quantum particles?

A new theory of everything, possibly with mathematical models, e.g., mathematical models of consciousness and creation or evolution, for the mathematically inclined who favor them, should scientifically supply the answers to some or all of the above questions. This is important as it would provide greater insight of our origin and nature.

14 SCIENTIFIC METHOD

It is the task of the theoretical physicist to explore feasible ideas and produce hypotheses without spending time in the laboratory. On the other hand, the experimental physicist derives his hypotheses from physical experiments - he works in the laboratory.

The theoretical physicist uses mathematical models to describe nature, and, since he has not had his ideas proven in the laboratory, they might be regarded with incredulity, especially if they were counter-intuitive or seemed to be against common sense, e.g., concepts such as time-travel and the existence of particles known as tachyons which travel faster than light (though it has been claimed that nothing could travel faster than light). Einstein had tried but had failed to produce a unified field theory which would explain why the four forces of nature, gravitational forces, weak nuclear forces, strong nuclear forces and electromagnetism, work together. The chapters here have been ambitious attempts at a better understanding of nature.

The theories, forecasts or speculations of theoretical physicists, which could be modeled with elegant mathematical equations that might be regarded as proofs of their validity, are still in need of physical confirmation. It is the experimental physicists who perform the experiments to confirm these theories, and, should the experimental evidences prove otherwise these theories would be discarded. No matter how sound the logic of these theories seemed to be, as detailed, e.g., in some elegant mathematical equations, they would be useless once they were contradicted by the experimental evidences.

Could the scientific community still insist on hanging on to the theories now and say that the experiments had been wrong and had thereby ended up with erroneous results, while insisting that the logic of the theories was so elegant that it could not be wrong? In other words, could they choose logic over empirical evidence? The answer is clearly "no". If they had so much faith in the logic, the experimental physicists would not be required. It would be contradictory, silly in getting the experimental physicists to obtain the empirical evidences and when they are found to contradict the theories, reject them. When it comes to the crunch, which is more important - elegant logic or contradictory empirical evidence? Of course, the empirical evidence should take precedence, as the physical senses are apparently "fool-proof" as compared to the logical faculty. The fact that experimental physicists are needed is a telling sign for logic, in this case, scientific logic. That is, logic is not that trustworthy or reliable, and, empirical fact is more reliable and important than logic.

But, when and where the empirical facts are not obtainable, logic would come in handy, i.e., we would have to resort to logical deduction to arrive at the truths. Arriving at the truths through physical experiment would apparently be the preferred way, provided it is feasible.

Is this not telling: for example, we tend to argue about the ethics (right or wrong) of things, but, hardly do we doubt or dispute what we or others see with our eyes, for instance, colors such as red, blue and green?

Hence, the importance of empirical evidence in science, as well as other areas of human activities, an importance which could be regarded as greater than that of logic.

15 THE IMPORTANT SOFT SCIENCE: ECONOMICS

Economics has to do with how buyers and sellers behave in the market-place, and, since most countries have free economies, there is generally little control over how buyers and sellers in the market-place behave. We could hope to influence them and change their behaviors through advertisements, incentives, tax policies, et al., but this is no guarantee that the desired change for the better would be forth-coming. Commercial logic might just fail to convince or change their minds. The three following chapters have some important suggestions on improving the world economic situation.

Economics is effectually the study of how people behave in the business environment. Mathematical models might be used to describe the interaction of buyers and sellers in the market-place, and, forecasting techniques might be utilized to foretell how buyers and sellers would interact under varying conditions. But, the fact is that all human beings tend to be whimsical and their behaviors in commerce and in every other thing would thus tend to be difficult, if not impossible, to predict. How many times economic forecasts are inaccurate and economic policies fail? Game theory, which had been pioneered by John Von Neumann, a mathematician, and for which John Nash, the mathematical genius of "Beautiful Mind" fame, won the Nobel Prize for economics, has been used as a model of competitive behavior in business. But, it is based on the "cooperative" concept, whereby groups of businesses might team together or group together for mutual benefits to compete with the others. This appears simplistic, as in the real situation businesses might not cooperate at all but would instead be all out to cut each other's throats.

16 ECONOMICS AS A SCIENCE FOR SOLVING PROBLEMS

Economics can be defined as the science of creating wealth. It is the theoretical underpinning which explains the workings of commerce. The important point to note is that despite the analyses and recommendations of the economic experts who advise, guide and formulate governmental economic policies, the dynamics of the global economy has been as intractable as ever. Governmental policies have always been bandied about to ensure that businesses thrive, people have jobs and the economy is buoyant, but often the reverse results. What can be really seriously affecting the economy? One may wonder.

It should be borne in mind that economics is about the thinking of people involved in commerce - whether they are buying, selling, employing, looking for jobs, trading, negotiating, investing, getting bank loans, lending, etc. Companies may reduce prices, offer free gifts, increase salaries, provide all kinds of incentives, etc., to attract customers and employees, customers may selectively purchase goods based on brand-name, product image, price, service, or a combination of all these, and workers may choose jobs and employers. To modulate all these commercial activities to ensure that there is prosperity and full employment in the economy, the government has an important, central role. The government, through the central bank, can increase or decrease bank interest rates to encourage or discourage savings and hence reduce or increase the money supply, and, also increase or decrease taxes to reduce or increase the money supply. (The government influences the economy through monetary policies, which pertain to the regulation of money supply, and, fiscal policies, e.g., increasing or lowering taxes and increasing or decreasing spending on public works.)

However, what many seem not to be well aware of is that varying the interest rates and taxation rates by the government quite often do not have the desired effects - increasing or decreasing the money supply, and, increasing or decreasing spending. To expect such governmental economic policies to work all the time is naïve. Such economic policies may work sometimes, but certainly not all the time. To understand why this is so, we need to have a deeper understanding of the human psyche.

Economics is actually ultimately about how people think and behave where money or wealth is concerned. There are many wealthy people who spend relatively little, and even hoard, despite their wealth, while in economic theory we assume that people with more money will spend more. On the other hand, contrary to economic theory, many not so well-off people spend relatively much despite their lack of wealth, and many may even beg, borrow or steal in order to afford a spendthrift, luxurious life-style, e.g., the shopaholics, the nightclubbers, etc. Though we may expect those with good incomes to save more when bank interest rates are high, this may not happen when the person is spendthrift and generous, e.g., interested in shopping, traveling, dining in fine style, clubbing and giving treats. Some of these people may have other financial commitments, e.g., mortgages to pay, children's education to finance, loans to repay, investments or other business undertakings, etc.; so they may not be able to save despite the high bank interest rates. Of course, those with lower incomes, who are less able to make ends meet, may also not be inclined or able to save even if the bank interest rates were very high.

High import duties and taxes may also not coerce people to spend less. For instance, despite the high import duties, road taxes and ERP charges for cars to discourage car-ownership, making car-ownership in the country really expensive, many who can ill afford to own cars possess them, for the love of the automobile, convenience and/or the sake of looking good (status symbol). Despite all the best efforts of the government here to reduce the car population in order to solve the problem of road congestion, car-ownership seems to be getting more robust. And, despite the high import duties and hence high prices of liquor and cigarettes here, those addicted to them evidently continue consuming them as before. We should also not under-estimate the effect of advertising or marketing gimmicks, which can be subtle but can cause impulse buying, especially in the case of consumer products such as food, drinks and clothing - here people who purchase are governed more by emotion or feeling than reason and may do so whimsically.

Also, do not be surprised that often the not so well-off will spend more than the rich. It must be noted that immaterial of whether they are wealthy or poor some people will simply spend, as it is their nature to be spendthrift and generous with money. Some people may just lack financial management skills and discipline where budgeting is concerned while some others are prudent and wise with their money. On the other hand, many wealthy people are miserly or "Jews" as they may be derisively called. That is, the wealthy, who are expected to spend more, or, save more, or, invest, may not do so, while many who are not well-off may spend a great deal, some even borrowing or stealing to do so.

There are also well-to-do people who think far ahead, plan and save for a great future or for the "rainy day" instead of spending freely. There are also people with entrepreneurial ambitions who may save in order to start a business, regardless of the bank interest rates and taxation rates. There are people who live simple lives, who will save a substantial part of their income because they have no interest in shopping, fine-dining, clubbing, traveling and other luxuries.

Doesn't all this explain the frequent failures of governmental monetary and taxation policies where the economy is concerned? Too little money circulating, i.e., too little spending, in the economy leads to depression and unemployment. Too much money circulating, i.e., too much spending, leads to inflation or high prices and financial hardship. However, there is an evident solution to all this. To solve the problem of depression, people can be "forced" to spend in order to increase the money supply and buoy up the economy. Here, two types of currency can be introduced, one type with expiry dates, which may be varied, to ensure that a certain quantity of money circulates during a certain period (these money have to be spent by certain dates and cannot be saved), the other type with no expiry date which can be spent anytime or saved. The expiry dates of the first type of currency can be varied as follows: short, medium or long term in expiry. The expiry dates can be adjusted from time to time according to the economic conditions. For example, if there is inflation, i.e., too much money is in circulation, the currencies with expiry dates may, for instance, specify that the money cannot be spent for the next few months/years (period), after which period they can be spent but will expire on certain dates. By thus playing around with the usage and expiry dates of these currencies, the amount of money, and spending, in the economy can be controlled. On the other hand, to counter the effects of deflation, wherein the quantity of money in circulation and spending are low, the expiry dates of these currencies can be shortened. However, in this instance, to avert the onset of inflation the production of goods and services should keep up with the increased demand caused by the increase in money supply and spending.

All this should be administered by a statutory body or governmental body. Employees and sellers will thus be paid two types of currencies, one with the above-mentioned expiry dates and the other without. What amount or proportion of the two types of currency to be paid out should be determined and administered by this statutory body or governmental body, based on the economic conditions and/or forecasts for the specific period. Neater still if instead of paying by issuing these two types of hard currency a debit card (payment based on the actual amounts of the two types of currency available in the person's debit card account, which will be deducted from the account when the account holder is making payments) is used; this debit card or cashless paying method should make it easier for the governmental institution to administer the whole monetary system. This will apparently be an effective way to overcome our on-going economic malaise over which we seem to have little or no control - depression, unemployment, inflation, deflation. Here we are only talking about monetary transactions within the same country.

At the international level, concerning trading between countries, the same principle should also apply, except that the currencies with expiry dates, the currencies without expiry dates and the debit cards should now be issued and administered by an international authority, e.g., a newly created division of the United Nations.

Summing up on the point why economics does not solve economic problems, it can be said that economics fail because people's minds, attitudes and behaviors are evidently very difficult, if not impossible, to predict. As is described above, they are complex and come in a great variety of types and characters, often emotional and not very rational, and full of whims and fancies which are quickly changing as well as habits, and therefore cannot be expected to respond appropriately to governmental economic policies which are based on assumptions that may be too simplistic and unrealistic. However, the monetary system described above should perform the trick.

In any case, we should control the economy and not let the economy control us.

17 SOLUTIONS IN ECONOMICS

Economic problems that plague us seem intractable and seem to require some radical countermeasures. One of the root causes of a recession is that people are not spending sufficiently. To overcome this problem a radical change in our monetary system could be effected.

Proposed Monetary System
The world economy has been frequently gloomy and depressed.

The economy works through the market forces of supply and demand. The level of demand ultimately determines the quantity of goods and services that will be made available in the market, though sometimes the quantity of goods and services produced in the market are so produced in anticipation of demand.

The crux of the matter is that money held by people has to be spent or change hands in sufficiently large volumes for the economy to be buoyant. If people lose their jobs, have little or no money to spend and curtail or stop spending the economy will be in trouble. Equally bad is the situation where employment rate is high, people have money to spend but are hardly spending. To rectify the situation of low employment governments can help to create jobs. But after the jobs have been created and people have jobs and money but are still not spending sufficiently, what will become of the economy? This is an interesting point to ponder. The author's view is that companies or producers of goods and services will flounder when their goods and services are not in demand as people have little inclination to spend the money they possess (unless the government now performs the charitable act of financially supporting them to ensure that their employees do not lose their jobs). When these companies or business units flounder, they will retrench or wind up, that is, those with jobs will now find themselves without jobs and income, and, spending will become less. So, it is more or less back to "square one". Of course in the situation whereby people have money but are not spending enough but saving the bulk of their income, the government may well be urged to resort to suitable fiscal policies to rectify the situation such as reducing bank rates and reducing taxes. Companies and business enterprises may well be advised to advertise and market their goods and services more aggressively to create or increase the demand for them. Will these be certain to change people's attitude towards spending? The certainty could hardly be expected.

The point is that governmental policies and aggressive advertising and marketing can only help to coerce and persuade people to spend more but the final choice of whether to spend or not still belongs to the people. In other words, the "customer is king", the market belongs to the buyer. Put bluntly, business, the economy (even the government), is at the mercy of the buyer or customer. Can we force the buyer or customer to spend? No? The author's answer is yes. But, a radical change has to be implemented.

All these suggest that the economy is indeed difficult, even impossible, to control. We should not allow the whims and fancies of the buyer or market to create problems in the economy, problems for businesses who need a profit to carry on, problems for people who need jobs, problems for the government whose responsibility is to see that their people have jobs and incomes. As mentioned earlier, creating jobs and incomes for the people may be fine and good, but when those with incomes are not spending sufficiently, it is not good for the economy, it is not good for everybody. We have to find a way to force everyone to spend sufficiently so that economically everyone will survive. The question here is how? The author would like to make a radical suggestion. The suggestion is that two types of currencies be introduced to ensure that there is sufficient spending. One type of currency has an expiry date while the second type of currency has no expiry date like our current ones. The issuance and administration of such currencies, especially the first type which has an expiry date, are to be the responsibility of a governmental monetary authority or statutory board. For example, the government can pass a law to the effect that all salaries have to paid in the following format: 70% in currencies which expire say three months from the date of payment and the balance 30% in currencies with no expiry date like our present currencies. The seller who receives the currencies with expiry dates from the buyer could then exchange them at the governmental monetary authority for the following currencies: 70% in currencies which expire in three months' time and 30% in currencies with no expiry date. In this way everyone, bosses and employees, and, buyers and sellers alike, have to spend at least 70% of their earnings within the allotted time frame of three months. Here we are only talking about salaries and purchases within the same country.

At the international level, where business between countries is concerned, the same principle should also apply, except that the currencies with expiry dates and the currencies without expiry dates should now be issued and administered by an international authority, for example, a newly created division of the United Nations.

The question is whether this arrangement is practical. Except for some teething problems on initial implementation, the author does not think implementation is going to be difficult. Though it entails some administrative costs, the price is worth paying. For sure, implementing this new monetary system will result in new jobs being created, which is good for the economy. And, most important of all, this new monetary system will help to eradicate recessions, unemployment and other economic and related social problems.

There is no point grumbling about the poor state of the economy and how miserable life is because one has no job and no income, no business and no earnings. Everyone, especially the powers that be, should seriously attempt to analyze what causes the economy to fail and people to lose their livelihood and try hard to do something about it.

Hence, the authorities should seriously consider the author's reasoned proposal here.

18 MORE SOLUTIONS IN ECONOMICS

In this chapter, some solutions to our modern seemingly intractable economic problems are proposed. Keynesian theories, the bug-bear of modern economics, seem to hold little water nowadays. Modern economies appear to be in need of more radical solutions.

Proposed Economic Solutions

We frequently suffer the throes of economic recession, becoming the victim of the economic recession. We hope, pray, and wait helplessly for the economy to recover. Governments often try to do something that may help the economy to recover fast, but there is no guarantee that they can help bring about a reasonably quick recovery. They may introduce more public works to create jobs, implement fiscal and monetary policies, re-train retrenched workers and carry out a myriad of other actions. But the economy may take its own time to pick up. The economy seems to live a life of its own, a life that does not seem easy to control at all. In fact, the fate of the economy is the fate of all human beings. Difficulty in controlling the economy simply translates to difficulty in controlling the human destiny.

In a depressed economy, human miseries are all too obvious. The jobless are living in fear and anxiety. Many have even turned to crimes (hence, the higher crime rates may be evident), such as robbery and looting. For those who are less well off, they now have to tighten their belts, live on a smaller budget. A number might seek the assistance of the authorities. It appears a helpless situation.

From the study of the history of past economic trends, everyone is more or less telling himself that when the economy swings downwards it can be expected to swing upwards again some time in the future (perhaps, in a few years' time), and, vice versa. That is, we assume that "history repeats itself". What, if in the future, history does not repeat itself? This is not an impossible scenario. It is indeed dicey.

This chaotic state of affairs which is not new and which brings about much uncertainty and misery seems to imply something. It seems to tell us that there is something wrong with our social system. It seems to imply at least some of the following, though they are perhaps not obvious to many as they are very complex indeed:-

1) There may be a problem in the distribution of wealth. The gap between the haves and the have-nots may have become very wide.
2) People may have been "psyched" by adverse publicity into hoarding rather than spending as usual, causing a fall in demand for goods and services all round, hence, reduction in production, retrenchment and loss of jobs.
3) Too much competition and under-cutting result in loss of profitability, hence loss of incentive to

produce or increase production, and, hence, business closure, retrenchment and loss of jobs. (Competition may not be really good, though consumers tend to deem it desirable.)

4) The free enterprise system, or, the capitalistic system, without much governmental control, may not be that efficient and good after all as it implies freedom for the economic system to be well beyond proper control which enables it to move along the correct, desirable path, ending up in the wrong, undesirable direction instead.

5) The wealthy has too much power and control over the workers and the have-nots. (In a recession, the workers and the have-nots will obviously suffer more than the wealthy, who may not hesitate to displace them from their jobs to save their own skins.)

6) The bosses exploit the workers, bringing misery to the latter one way or another. Though the labor legislations protect the workers and ensure that their minimal entitlements are fulfilled, bosses can still find subtle means to exploit their workers. (This is perhaps one of the greatest flaws of the capitalist system - exploitation, greed, or, profit-maximization at the expense of the workers.)

7) The bosses may also exercise some form of subtle control over the government, e.g., by sponsoring politicians (who run the government) or threatening to re-locate to another country, thus possibly causing job loss and a poor "employment situation" (a political headache for an elected government).

8) It may be a situation that even those who are desperate for a job and are prepared to accept a low salary may have hardly any opportunity of employment. The government cannot force companies to hire anyone they do not wish to hire. (In this respect, a socialist economy has an advantage in that its government-controlled industries can provide plenty of job opportunities for the populace.)

9) Instead of being the master of technology, we may have become its slave or victim. Technology can be changing so rapidly that it is very difficult or impossible to keep up with it.

10) The economy of one nation may have been adversely affected by the economies of the others and vice versa.

We, especially our governments, should manage the economy, instead of allowing the economy to manage us. We have to re-think and re-vamp our economic policies on a global scale, with nations consulting one another. It appears that capitalism is far from perfect as an economic or social system as it encourages human greed and exploitation. It may be better to combine the desirable elements of both the capitalist and socialist systems. Otherwise, we may have to reformulate a new social/economic system, which is radically different from both the capitalist and socialist systems. How about a world association to manage and co-ordinate the economies and economic activities of all the nations in the world, this association being represented by members from all the nations in the world? This association can fix or set quotas for imports and exports amongst all the world's nations. For example, the association can declare that Country A exports Item X to Country B and Item Y to Country C and imports Item S from Country B and Item T from Country C, Country D exports Item J to Countries E, F, G and H and imports Item K from Country E, Item L from Country F, Item M from Country G and Item N from Country H, et al., with the quantities imported and exported amongst them stipulated, subject to periodic changes in the future. In this way all countries in the world will have to produce sufficient goods and services to fulfill their own consumption needs as well as meet the quotas set by the association. Thus, there should be enough jobs and incomes for everyone.

Contrary to general opinion, competition is really not good for business, especially in the long run. As mentioned in (3) above, too much competition and under-cutting may lead to business closure, retrenchment and loss of jobs, which is indeed not good for everybody. Fair enough, competition ensures that the consumer does not have to pay a high price for his good or service.

Also, it is felt that competition will encourage businesses to be more productive (for example, to be more competitive or better businesses will strive to be more efficient and more cost-conscious, but they may be so cost-conscious that the welfare and safety of their workers are compromised). Cooperation and team-work amongst competitive businesses are better for there are economies of scale (resulting in cost-savings and thus, theoretically at least, lower prices for the consumer) and more heads to do the thinking and thrash ideas. However, to protect the consumer from over-priced goods or services, the government should have the right to step in and regulate prices.

Finally, a social or economic system should consider the general welfare and economic well-being or safety of every human being in the world. What we are having now is social and economic imbalance, with some people and nations having great wealth, some not so great wealth, and some having abject poverty (with problems such as famine, undernourishment and disease). Economic problems tend to lead to social problems such as crimes, riots, looting or even rebellion. The economic system should take into consideration the welfare of every human being and should not have allowed some to wallow in vast wealth while some others are down in the economic dumps. Our social or economic system should ensure a more even distribution of wealth. A more even distribution of wealth will lead to greater social equanimity.

19 CONCLUSION

There are many things in life that still puzzle us. First and foremost, there is the question of consciousness. Consciousness is apparently the result of electrical messages passing through the brain. Moreover, to be conscious a person has to be physically alive, i.e., his heart has to pump blood through his body, blood has to circulate through his body and his lungs have to inhale and exhale. Consciousness is awareness or alertness and is intangible. It could be regarded as the spirit or soul of a person and is mysterious. Consciousness or mind apparently has influence or control over matter, over our environment. Where was a person's consciousness or spirit before he was born into the world and how did it come about when he was born into the world? Would consciousness or spirit be able to survive a person's death? Consciousness represents the self, humanity itself, but it is apparently still as intriguing and mysterious as ever. Though scientists might probe the mysteries of nature they have, strangely, not been able to plumb the deep mysteries of consciousness, their own consciousness even, and the afore-mentioned questions remain unanswered, or, unanswerable.

Next, the subject of gravity is apparently still a puzzle to modern-day scientists. It is odd that though the effects of gravity are evident in the macroscopic cosmos they appear to be absent in the microscopic quantum world (where quantum particles whiz about with great abandon without displaying the effects of gravity). It appears that the physical laws of the cosmos differ from those of the quantum world. How come? The scientists are apparently stumped. There are several theories relating to gravity, e.g., the theory that gravitons (particles) and gravity waves exist, which, hopefully, would be detected in experiments. Gravity is now regarded as an attractive force, but it is not the attractive force of positive and negative particles, a phenomenon known as magnetism. Perhaps, it is an attractive or pulling force in reverse, a pushing force. The sun apparently exerts a gravitational attraction on the earth, as well as the other planets, which all orbit round it. This attraction is apparently stable, for otherwise the planets would have been displaced from their respective orbits and collide with each other. Many apparently regard this efficient system of planetary movements as the work of God, the result of willful, intelligent design. Instead of regarding gravity as a force in the earth's core that pulls everything towards it, how about considering it as an outward force that pushes everything towards the center of the earth? Is this not also a plausible theory? This pushing force could originate from the space surrounding us, e.g., the air, gases, light, et al. Light and electromagnetism have been found to be the same thing. Perhaps, gravity may be similar to some other aspect or facet of nature.

Finally, the existence of perpetual motion machines has been debunked. But are not the various planets in our universe and other universes perpetual motion machines per se, as they apparently orbit in their various paths non-stop and perpetually for innumerable years? What is the driving force behind their apparent perpetual motions? Is this the handiwork of a higher Being or Intelligence?

It would be great to have all these matters sorted out, to have their logic revealed.

BIBLIOGRAPHY

[1] F. David Peat, Superstrings And The Search For The Theory Of Everything, Contemporary Books, 1988

[2] John Gribbin, The Search For Superstrings, Symmetry, And The Theory Of Everything, Back Bay Books, 2000

[3] Steven Weinberg, Dreams Of A Final Theory (The Search For The Fundamental Laws Of Nature), Vintage, 1993

[4] Paul Davies, Superforce (The Search For A Grand Unified Theory Of Nature) Counterpoint, 1984)

[5] John Gribbin, In Search Of Schrodinger's Cat (Quantum Physics And Reality), Bantam Book, 1984

[6] P. A. M. Dirac, The Principles Of Quantum Mechanics, Oxford University Press, 1958

[7] P. A. M. Dirac, The Evolution Of The Physicist's Picture Of Nature, Scientific American, May 1963

[8] J. B. Hartle & S. W. Hawking, The Wave Function Of The Universe, Physical Review, Vol. D28 (1983)

[9] David Lindorff, Pauli And Jung: The Meeting Of Two Great Minds On The Unity Of Matter And Spirit, Quest Books, 2004

[10] Roger Penrose, The Emperor's New Mind: Concerning Computers, Minds, And The Laws Of Physics, Oxford University Press, 2002

[11] Roger Penrose, Shadows Of The Mind: A Search For The Missing Science Of Consciousness, Oxford University Press, 1996

[12] James Lovelock, Gaia: A New Look At Life On Earth, Oxford University Press, 2000

[13] David Bohm, Wholeness And The Implicate Order, Routledge, 1980

[14] E. Merzbacher, Quantum Mechanics, Second Edition, John Wiley & Sons, 1970

[15] Marcus Chown, The Quantum Zoo: A Tourist's Guide To The Neverending Universe, Joseph Henry Press, 2006

[16] A. Vilenkin, Predictions From Quantum Cosmology, Physical Review Letters, Vol. 74, 1995

[17] A. Modinos, Quantum Theory Of Matter, John Wiley & Sons, 1996

[18] Rupert Sheldrake, A New Science Of Life: The Hypothesis Of Morphic Resonance, Park Street Press, 1995

[19] A. Beiser, Concepts Of Modern Physics, Fifth Edition, McGraw-Hill, 1995

[20] Brian Green, The Elegant Universe, Vintage Books, 2000

[21] John Maynard Keynes, The General Theory Of Employment, Interest, And Money, Harcourt, Inc., 1964

[22] Mark Skousen, The Making Of Modern Economics, M. E. Sharpe, 2009

www.ingramcontent.com/pod-product-compliance
Lightning Source LLC
Chambersburg PA
CBHW081735170526
45167CB00009B/3831